Enzymes and Drugs

Enzymes and Drugs

Serge Kirkiacharian
Julien Dumond

iSTE

WILEY

First published 2024 in Great Britain and the United States by ISTE Ltd and John Wiley & Sons, Inc.

ISTE Ltd
27-37 St George's Road
London SW19 4EU
UK

www.iste.co.uk

John Wiley & Sons, Inc.
111 River Street
Hoboken, NJ 07030
USA

www.wiley.com

Library of Congress Control Number: 2024933510

British Library Cataloguing-in-Publication Data
A CIP record for this book is available from the British Library
ISBN 978-1-78630-939-6

Contents

Preface

The development of new drugs is a multidisciplinary activity. It draws on various fields such as chemistry, physics, computer science and galenics, as well as proteomics, biology, physiology and enzymology, which play a major role.

The objectives of this book are to:

– help biochemistry and biology students in the faculties of medicine, pharmacy, veterinary medicine, engineering and medicinal chemistry to gain access to the field of medicinal enzymology;

– make up-to-date information on this discipline available to professional biochemists, chemists, biologists, physicians, pharmacists and veterinarians;

– provide useful documentation for basic and applied research in pharmaceutical companies, hospitals and universities;

– highlight the many therapeutic applications linked to enzymes, with the help of various examples;

– foresee the considerable potential of this discipline for the discovery of new drugs with enzyme activators, inhibitors and enzyme substitutes in the case of many rare diseases.

Despite our meticulous attention to the text we have prepared and checked, errors may have occurred. As is customary, the authors and publisher accept no responsibility for any consequences that may result, and thank readers in advance for informing them of any such errors.

March 2024

Preface

March 2024

Introduction

History

Although enzymes have been involved in fermentation processes since late prehistoric times, it was only in the 18th century that the first notions of "pure" enzymology appeared, with the experiments of Lazzaro Spallanzani. In order to investigate whether digestion was more than just a physical process, he performed several protocols to demonstrate the intervention of other biochemical factors. He was unable to qualify these factors, but he had demonstrated that they existed. These included digestion experiments following the ingestion of pierced wooden capsules filled with meat, as well as tests conducted with tubes containing the gastric juices of birds with meat or cereals, placed under the armpits.

Later, in the 1830s, the first enzyme was characterized. Its role was the catalysis of starch into maltose. The term "diastase", from the Greek *diastasis*, was first attributed to it and later replaced by "enzyme", from the German word *Enzym*, of Greek origin *En* ("in") and *Zumé* ("brewer's yeast").

In 1897, the Buchner brothers discovered the enzymes involved in the alcoholic fermentation process, initially in a therapeutic context, and proved that they function under specific in vitro conditions.

In 1922, the protein nature of enzymes was established, and in 1926, urease, the first enzyme, was isolated in its purest form and crystallized by James Sumner.

By the 1940s, hundreds of enzymes had been purified and crystallized. However, the first crystallographic structure of a protein deduced from X-ray diffraction (Kendrew et al. 1958) (which was myoglobin), and the first amino acid sequencing of ribonuclease (White and Anfinsen 1959), were not obtained until some 20 years later.

Enzymes are therefore protein-based biological catalysts. Catalysts are chemical reaction accelerators. Some RNAs may also have catalytic activity (Altman and Cech 1989).

Life depends on structures, within which chemical reactions take place in an orderly manner in time and space (the anabolism and catabolism of living organisms). Without enzymes, chemical reactions would not be compatible with life. Their dysfunction results in the appearance of various pathologies, which are nowadays partially overcome by the development of new drugs thanks to scientific progress in many fields.

Fields of application

Enzymes were first studied for their involvement in the metabolism of living organisms. Initially, these were easily accessible molecules (derived from liquids or simple organisms). When purification processes were improved, many enzymes were isolated. Once characterized, their possible applications by and for humans could be considered. Today, there are three major areas of enzyme targeting and industrial applications:

– pharmaceutical and cosmetics companies;

– food companies (Kilara and Shahani 1979);

– environment-related companies with a view to generating environmentally friendly products (Baupai 1999; McKinlay et al. 2007).

Sometimes, the same product can be used in all three areas (as in the case of succinate or mannan (Zeikus et al. 1999; Singh et al. 2018)). It should also be noted that enzymatic reactions are used to carry out various transformations in organic synthesis and medicinal chemistry. However, this vast field will not be covered in this book. The same applies to enzymatic activity determinations used as a means of clinical diagnosis.

The objectives of this book

The progress made in enzymology since the birth of this science has made it possible to broaden its various fields of application.

The aim of this book is to highlight the importance of enzymes in medicinal chemistry, and to consider present and future strategies to overcome enzymatic deficiencies.

Each enzyme molecule has a well-defined mission within the cell, catalyzing a specific chemical reaction. It is an indispensable part of a whole that goes beyond it, but which desperately needs it. If it does not work, or does not work well enough, or even too well, the organism suffers the consequences. Strategies are then available to correct these transient or permanent failures.

The book is divided into five chapters. The first deals briefly with the catalytic activity of enzymes and the various techniques for studying enzymatic kinetics. Subsequent chapters focus on the role of enzymes in the various therapeutic fields of viruses, bacteria, fungi, protozoa and humans. They cover the following topics:

– the mechanism of action of various drugs;

– the development of treatments or substitute enzymes to correct their deficiency;

– the discovery of new enzyme targets with a view to developing new treatments through the contributions of medicinal chemistry and genetics.

It is worth noting that the pharmaceutical industry uses enzymes to produce intermediates and/or chiral drugs in a single step, offering significant economic and ecological advantages over conventional organic synthesis processes. However, this vast field will not be covered in this book.

References

Altman, S. and Cech, T.R. (1989). Propriétés catalytiques de certains ARN. Nobel Prize in Chemistry [Online]. Available at: https://www.nobelprize.org/prizes/chemistry/1989/summary/.

Bajpai, B. (1999). Application of enzymes in the pulp and paper industry. *Biotechnol. Prog.*, 15(2), 147–157.

Kendrew, J.C., Bodo, G., Dintzis, H.M., Parrish, R.G., Wyckoff, H., Phillips, D.C. (1958). A three-dimensional model of the myoglobin molecule obtained by x-ray analysis. *Nature*, 181(4610), 662–666.

Kilara, A. and Shahani, K.M. (1979). The use of immobilized enzymes in the food industry: A review. *CRC Crit. Rev. Food Sci. Nutr.*, 12(2), 161–198.

McKinlay, J.B., Vieille, C., Zeikus, J.G. (2007). Prospects for a bio-based succinate industry. *Appl. Microbiol. Biotechnol.*, 76(4), 727–740.

Singh, S., Singh, G., Arya, S.K. (2018). Mannans: An overview of properties and application in food products. *Int. J. Biol. Macromol.*, 119, 79–95.

White, F.H. and Anfinsen, C.B. (1959). Some relationships of structure to function in ribonuclease. *Ann. N Y Acad. Sci.*, 81, 515–523.

Zeikus, J.G., Jain, M.K., Elankovan, P. (1999). Biotecnology of succinic acid production and markets for derived industria products. *Appl. Environ. Microbiol.*, 51, 545–552.

1

Enzyme Kinetics, Inhibition and Activation

Biological reactions are essentially catalyzed by enzymes. Some enzymes function solely on the basis of their protein structure, in other words, the spatial arrangement of their amino acids. Others need to be coupled to an atom or molecule in order to act. These are enzymes linked to functional co-factors, which may be ionized atoms, small molecules or coenzymes.

In any chemical reaction where a reactant (R) must form a product (P), the R molecules must have enough stored energy to reach an activation state, also known as a transition state. The reaction rate constant k is related to the free energy of activation (Ea) under standard conditions by Arrhenius' semi-empirical law:

$$k = A \cdot e^{-(Ea/RT)}$$

where A is the Arrhenius constant, R is the ideal gas constant ($J.K^{-1}.mol^{-1}$) and T is the temperature in kelvin (K).

The velocity of a chemical reaction in which R is transformed into P is proportional to the concentration of R molecules in the transition state. The activation energy, at a given temperature, indicates the energy required to drive one mole of R to the transition state.

An enzymatic catalyst can be used to lower the activation energy and optimize the chemical reaction. As enzymes are biological macromolecules, they are sensitive to pH and temperature.

To monitor an enzymatic reaction in which a substrate (S) is transformed into a product (P), units of enzymatic activity are used. The most common unit is the amount of enzyme required to transform one micromole of substrate per minute at 25°C. Specific activity corresponds to enzyme activity per milligram of protein (access to the enzyme's degree of purity). Finally, specific molar activity is the amount of substrate transformed in one minute by one mole of enzyme.

The most useful and common kinetics observed in enzymology are briefly presented here. Specific kinetics obtained for complex enzymatic systems, such as the proteasome (Stein et al. 1996), are not discussed.

1.1. Michaelis and Menten theory (Michaelis and Menten 1913; Kenneth and Roger 2011)

Enzymatic kinetics is no exception to the rules of classical chemical kinetics. However, there is one specific feature to bear in mind: the limited concentration of the biological catalyst can be saturated by the reaction substrate.

By plotting v_i (initial reaction velocities) as a function of the different substrate concentrations used (graph v_i = f ([S]) (Figure 1.1)), it is possible to accurately determine the parameters of the Michaelis constant K_M and the maximum velocity V_{max}.

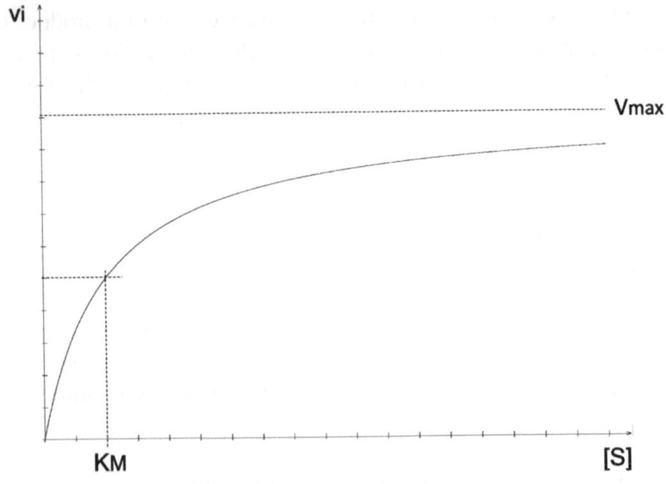

Figure 1.1. *Graph v_i = f ([S]) for at least 10 different substrate concentrations. A hyperbolic curve is obtained within the framework of Michaelian kinetics*

A hyperbolic curve is obtained within the framework of Michaelian kinetics. At low substrate concentrations, the reaction rate is proportional to the substrate concentration. Under these conditions, the reaction is pseudo-first order. Then, as the substrate concentration continues to rise, the reaction rate increases again, but no longer proportionally. Finally, when the enzyme is completely saturated with substrate, a plateau is reached. The enzyme is operating at full capacity, and the system is at V_{max}.

V_{max} is the maximum v_i obtained under the conditions of the experiment. The K_M (Michaelis constant) represents the substrate concentration for which $v_i = V_{max} / 2$. The K_M is used to approximate the affinity of the substrate for the enzyme.

The equation of the resulting equilateral hyperbola is:

$$v_i = V_{max} [S]/(K_M + [S])$$

An effector modifies the rate of a reaction by combining with the enzyme alone [E] and/or with the enzyme/substrate complex [ES]. Some effectors are inhibitors, slowing down the rate of reaction, while others are activators, speeding it up. This book is dedicated to effectors. These "classical" characterizations are extremely important for understanding the mechanism of action of molecules marketed for the treatment of disease. In 2002, 47% of therapeutic molecules targeted enzymes (Hopkins and Groom 2002). In 2011, the estimate was 29% (Rask-Andersen et al. 2011).

1.2. Irreversible inhibitors

These derivatives bind covalently to the active site. Once bound, the irreversible inhibitor renders the enzyme permanently non-functional by preventing substrate transformation.

The reaction is as follows:

$$E + I \underset{K_I}{\leftrightarrows} EI \underset{k_{inactivation}}{\rightarrow} E_i$$

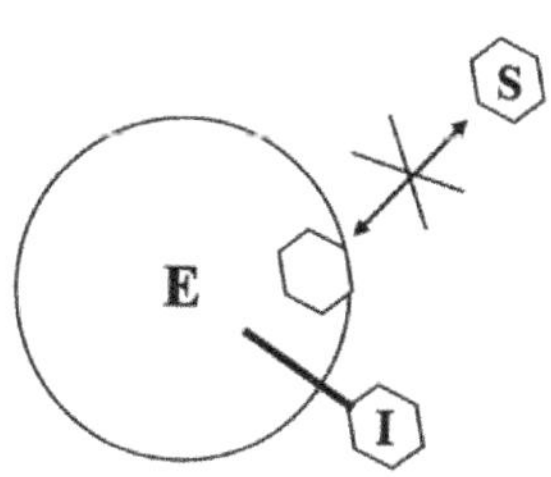

By plotting $1 / k_{observed}$, with the $k_{observed}$ deduced from assays carried out with different inhibitor concentrations, as a function of $1 / [I]$, Ki (inhibition constant) and $k_{inactivation}$ (kinetic constant of inactivation) can be deduced.

The standard equation of the line is:

$$1 / k_{observed} = 1 / k_{inactivation} + (K_i / k_{inactivation}) \cdot (1 / [I]) \text{ (Figure 1.2)}$$

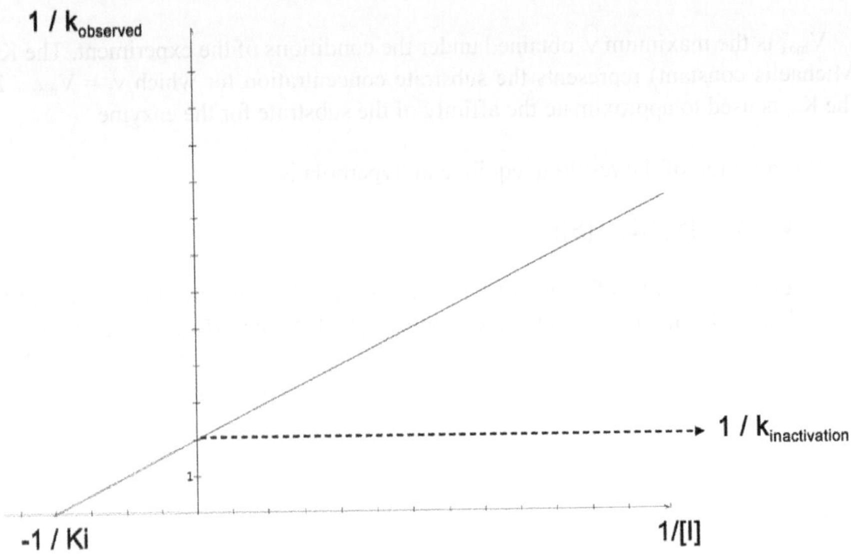

Figure 1.2. *Irreversible inhibition graph: 1 / $k_{observed}$ = f (1 / [I]) for at least three different inhibitor concentrations. A straight line is obtained. K_i and $k_{inactivation}$ can be deduced*

1.3. Reversible inhibitors in a Michaelian system

1.3.1. *Reversible competitive inhibitors*

These inhibitors (I) combine with the free enzyme (E) and compete with the substrate (S) at the enzyme's active site.

These inhibitors bind reversibly to the catalyst, forming an EI complex similar to the ES complex. The reaction is as follows:

In the presence of a competitive inhibitor, the Michaelis and Menten equation becomes:

$$v_i = V_{max} [S]/(K_M (1 + [I]/K_I) + [S])$$

k_{-1}

$E + S \quad \leftrightarrows \quad ES \quad \rightarrow \quad E + P$

$k_1 \qquad k_2$ (slow step)

$+ \qquad K_d\ (K_M)$

I

$K_i \quad k_i \downarrow\uparrow k_{-i}$

EI

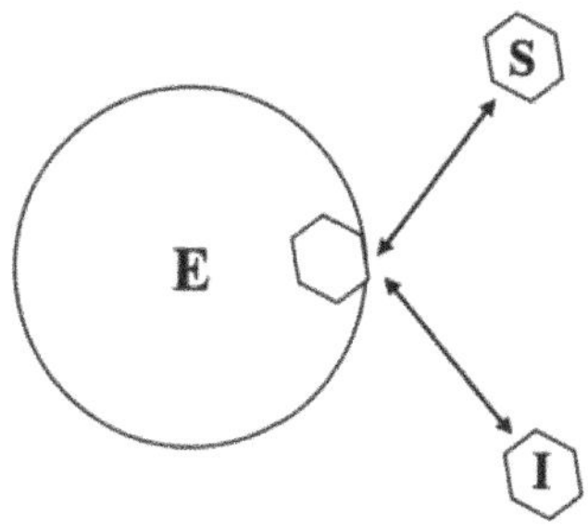

This is equivalent to having a constant K_M increased by a factor $(1 + [I] / K_I)$ without any change in V_{max} (Figure 1.3).

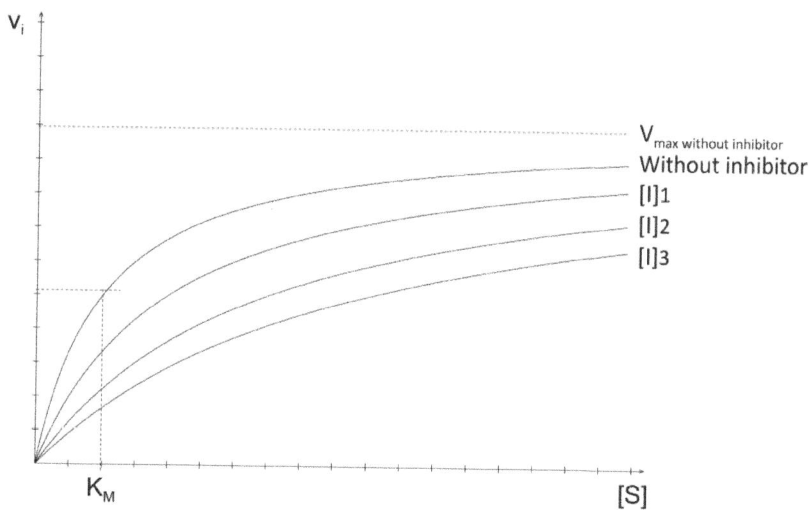

Figure 1.3. *Graph $v_i = f([S])$ obtained for an inhibitor-free control and three effector concentrations. Competitive inhibition*

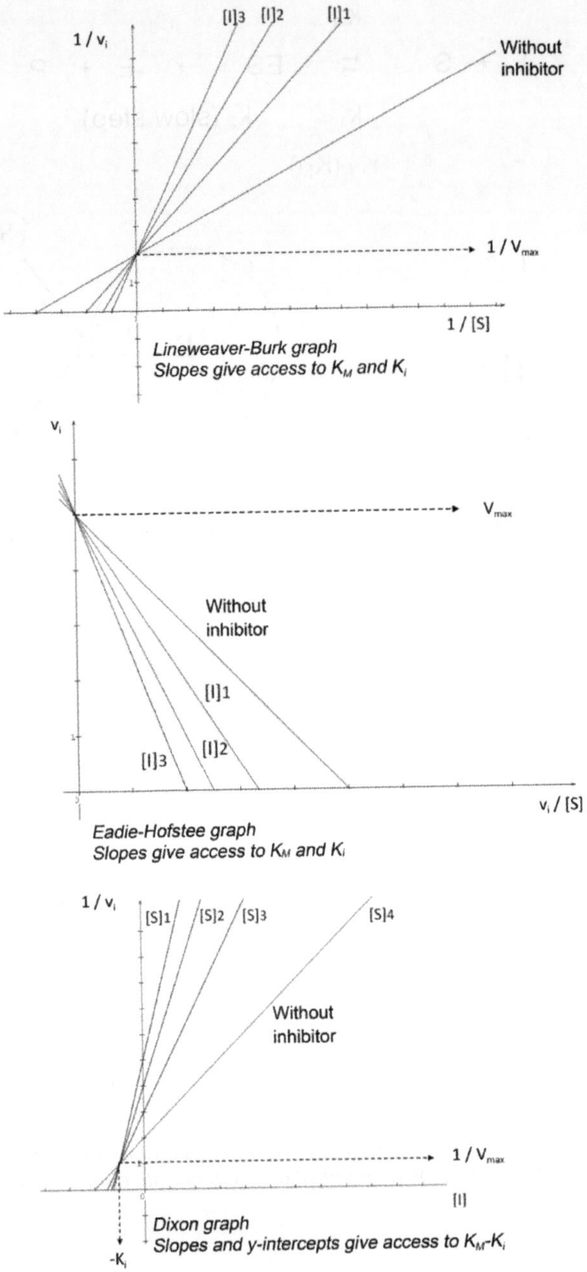

Figure 1.4. *Graphs by Lineweaver and Burk, Eadie Hofstee and Dixon. Competitive inhibition*

The Lineweaver and Burk, Eadie Hofstee and Dixon representations for a competitive inhibitor are shown in Figure 1.4.

The equations of the lines are shown below.

The particular characteristics of the lines for these graphs (competition (converging lines), slopes of the lines in each case) give precise access to the kinetic parameters (V_{max}, K_i and K_M).

– Equation of the straight lines obtained for the Lineweaver and Burk representation:

$$1/v_i = \frac{K_M(1+[I]/K_i)+[S]}{V_{max}[S]} = (K_M/V_{max}).(1+[I]/K_i).1/[S]+1/V_{max}$$

y *slope* *x* *y-intercept*

– Equation of the straight lines obtained for the Eadie Hofstee representation:

$$v_i = V_{max} - \{K_M(1+[I]/K_i)\}.v_i/[S]$$

y *y-intercept* *slope* *x*

– Equation of the lines obtained for the Dixon representation:

$$1/v_i = \frac{K_M(1+[I]/K_i)+[S]}{V_{max}[S]} = \{K_M/(K_i.V_{max}.[S])\}.[I]+(1+K_M/[S])/V_{max}$$

y *slope* *x* *y-intercept*

1.3.2. *Reversible non-competitive inhibitors*

These inhibitors (I) can bind to the free enzyme (E) or to the enzyme that has bound the substrate (ES complex).

These inhibitors combine reversibly with the catalyst outside its active site. The reaction is as follows:

$$
\begin{array}{ccccc}
 & k_{-1} & & & \\
E + S & \leftrightarrows & ES & \rightarrow & E + P \\
 & k_1 & & & k_2 \text{ (slow step)} \\
+ & K_d\ (K_M) & + & & \\
I & & I & & \\
K_i\ k_i \downarrow\uparrow k_{-i} & & K_{i'}\ k_{i'} \downarrow\uparrow k_{-i'} & & \\
EI + S & \leftrightarrows & ESI & &
\end{array}
$$

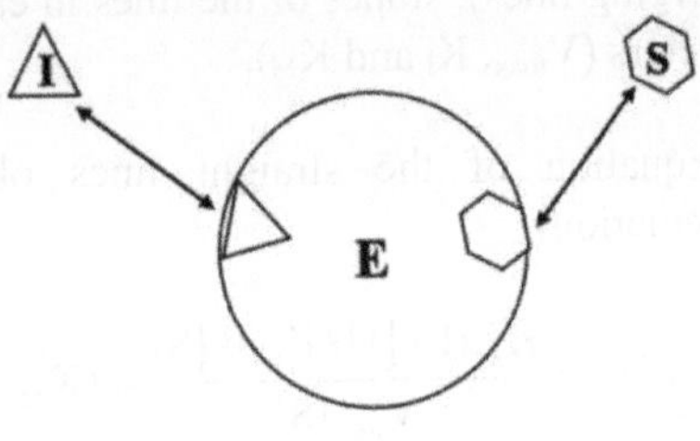

The equations of the curves are given below. The particular characteristics of the lines obtained for the relative graphs (parallelism, competition (converging lines), slopes of the lines according to each case) give precise access to the kinetic parameters (V_{max}, K_i and K_M).

– Equation of the straight lines obtained for the Lineweaver and Burk representation:

$$\underset{y}{1/v_i} = \frac{K_M + [S]}{(V_{max}/(1+[I]/K_i)).[S]} = \underset{slope}{(K_M/V_{max}).(1+[I]/K_i)}.\underset{x}{1/[S]} + \underset{y\text{-}intercept}{1/(V_{max}/(1+[I]/K_i))}$$

– Equation of the straight lines obtained for the Eadie Hofstee representation:

$$\underset{y}{v_i} = \underset{y\text{-}intercept}{V_{max}/(1+[I]/K_i)} - \underset{slope}{\{K_M(1+[I]/K_i)\}}.\underset{x}{v_i/[S]}$$

– Equation of the lines obtained for the Dixon representation:

$$\underset{y}{1/v_i} = \frac{K_M + [S]}{(V_{max}/(1+[I]/K_i)).[S]} = \underset{slope}{\{(1/(K_i.V_{max})).(1+(K_M/[S])\}}.\underset{x}{[I]} + \underset{y\text{-}intercept}{(1+K_M/[S])/V_{max}}$$

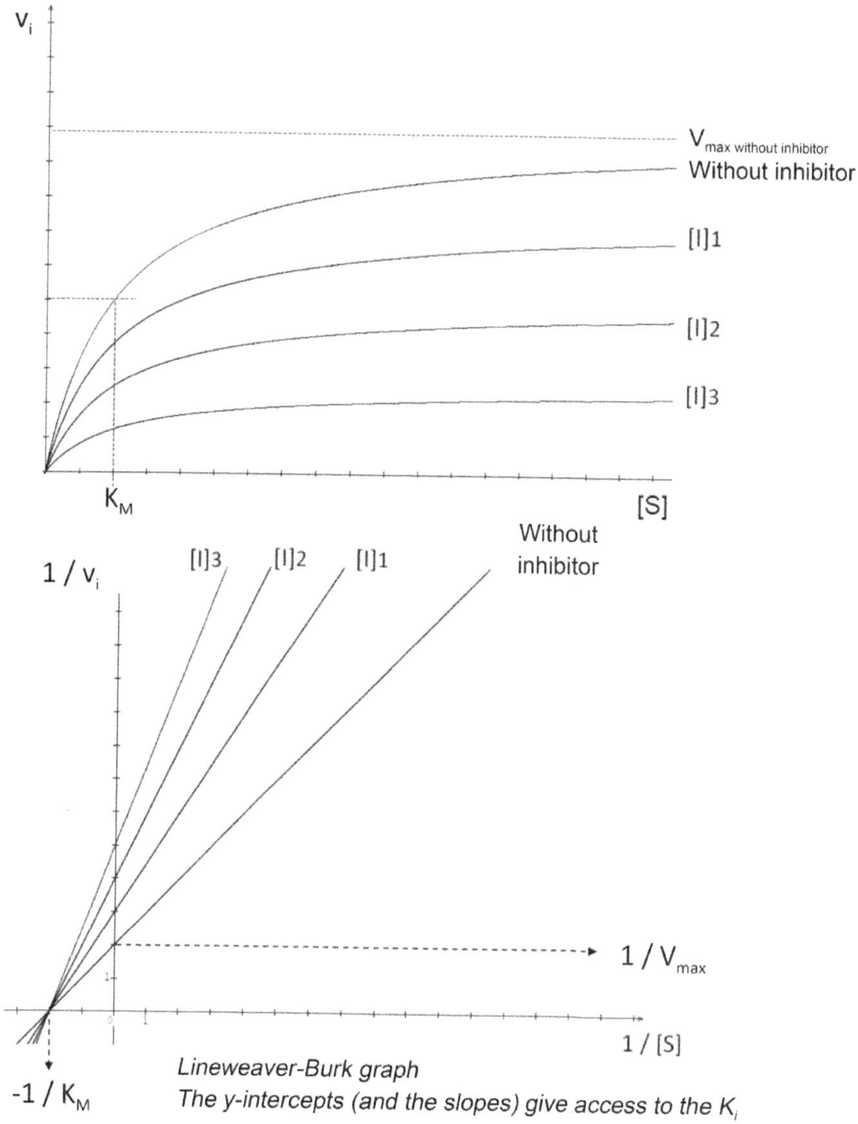

Figure 1.5A. *Graphs $v_i = f([S])$ and Lineweaver and Burk. Absolute non-competitive inhibition*

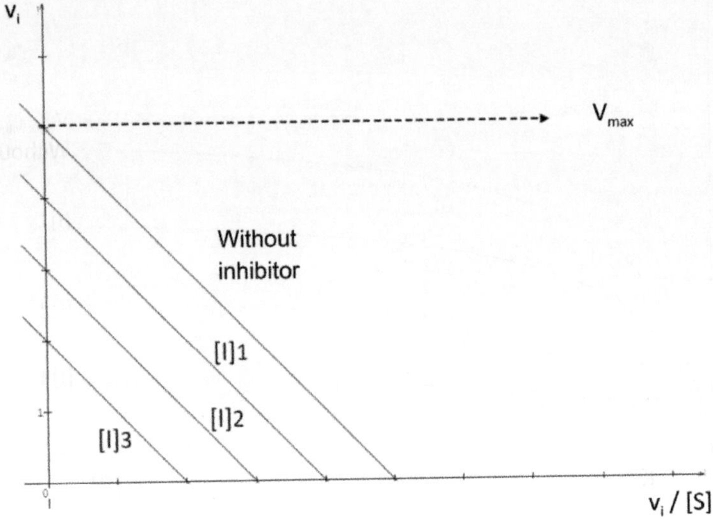

Eadie-Hofstee graph
The slopes give access to K_M and y-intercepts give access to K_i

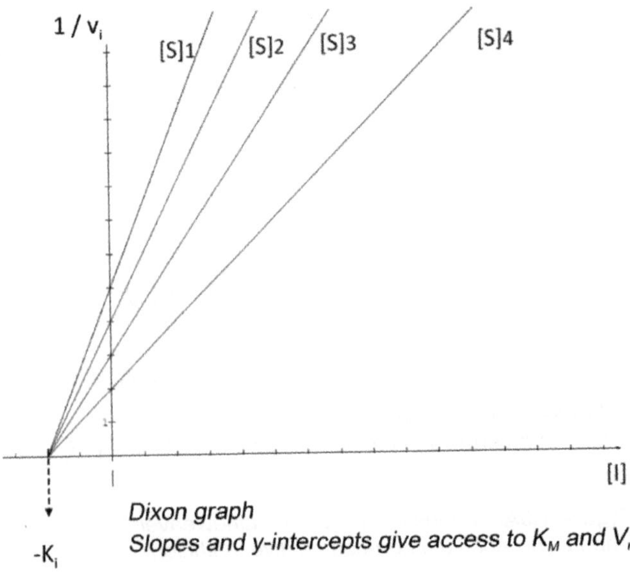

Dixon graph
Slopes and y-intercepts give access to K_M and V_{max}

Figure 1.5B. *Eadie Hofstee and Dixon representations. Absolute non-competitive inhibition*

Figure 1.5 shows the v_i = f ([S]) graph and the Lineweaver and Burk, Eadie Hofstee and Dixon representations for an absolute non-competitive inhibitor (which binds with the same affinity to the free enzyme and to the enzyme–substrate complex with $K_i = K_i$').

1.3.3. *Reversible uncompetitive inhibitors*

These inhibitors (I) only bind to the enzyme that has bound the substrate (ES complex). The result is an ESI complex. The inhibitor does not compete with the substrate, and its action requires prior binding of the substrate to the enzyme. A conformational change reveals the inhibitor's binding site. The reaction is as follows:

$$E + S \underset{k_1}{\overset{k_{-1}}{\leftrightarrows}} ES \xrightarrow[k_2 \text{ (slow step)}]{} E + P$$

K_d (K_M)

$$ES + I \underset{k_{i'}}{\overset{k_{-i'}}{\rightleftarrows}} ESI \quad K_{i'}$$

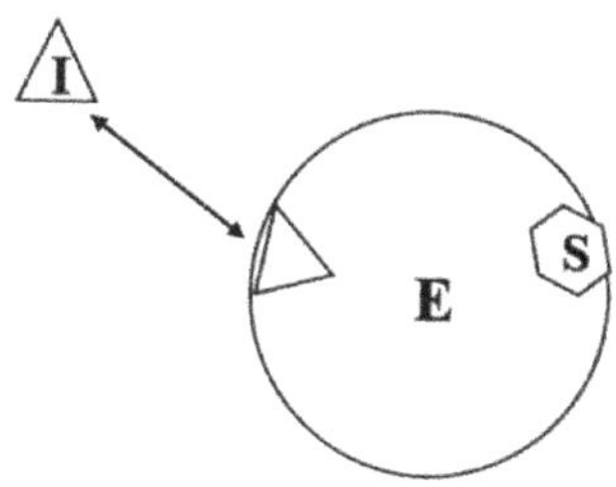

Figure 1.6 shows the v_i = f ([S]) graph and the Lineweaver and Burk, Eadie Hofstee and Dixon representations for an uncompetitive inhibitor. The equations of the curves are shown below. The particular characteristics of the lines obtained for the relative graphs (parallelism, competition (converging lines), slopes of the lines according to each case) give precise access to the kinetic parameters (V_{max}, K_i and K_M).

– Equation of the straight lines obtained for the Lineweaver and Burk representation:

$$1/v_i = \frac{K_M + [S](1+[I]/K_{i'})}{V_{max}[S]} = (K_M / V_{max}).1/[S] + 1/(V_{max}/(1+[I]/K_{i'}))$$

y *slope* *x* *y-intercept*

– Equation of the straight lines obtained for the Eadie Hofstee representation:

$$\underset{y}{v_i} = \underset{y\text{-intercept}}{V_{max}/(1+[I]/K_{i'})} - \underset{slope}{\{K_M(1+[I]/K_{i'})\}}.\underset{x}{v_i/[S]}$$

– Equation of the lines obtained for the Dixon representation:

$$\underset{y}{1/v_i} = \frac{K_M + [S](1+[I]/K_{i'})}{V_{max}\,[S]} = \underset{slope}{\{1/(K_{i'}.V_{max})\}}.\underset{x}{[I]} + \underset{y\text{-intercept}}{(1+K_M/[S])/V_{max}}$$

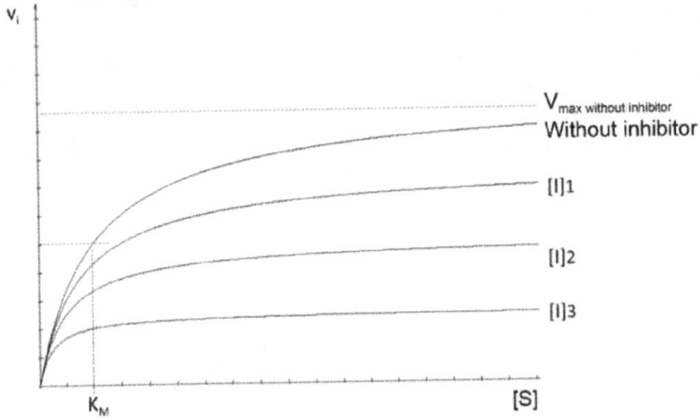

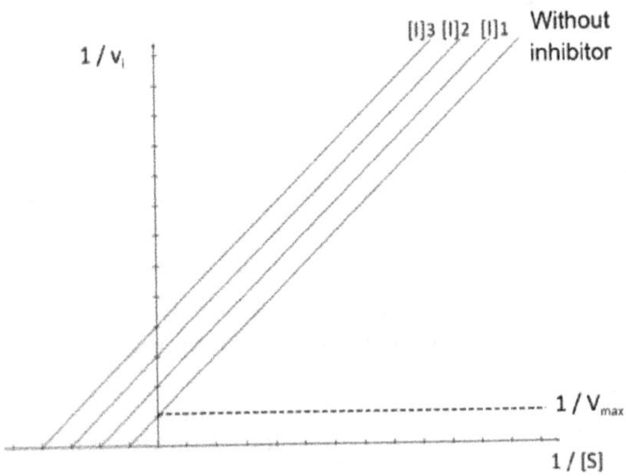

Lineweaver-Burk graph
The y-intercepts give access to the $K_{i'}$ and the slopes give access to the K_M

Figure 1.6A. *Graphs $v_i = f([S])$ and Lineweaver and Burk. Uncompetitive inhibition*

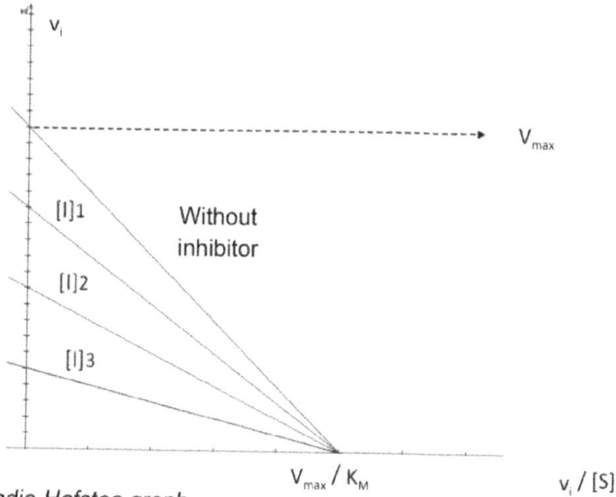

Eadie-Hofstee graph
The y-intercepts (and the slopes) give access to the K_i.

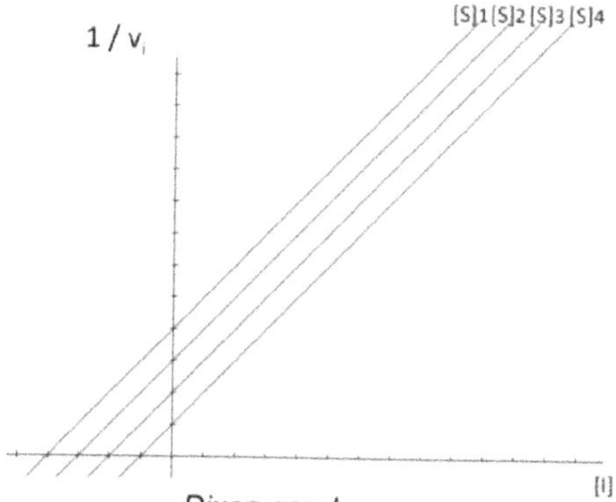

Dixon graph
Slopes give access to K_i.
Y-intercepts give access to K_M
Slopes and y-intercepts give access to V_{max}

Figure 1.6B. *Eadie Hofstee and Dixon representations. Uncompetitive inhibition (continued)*

1.4. Allostery: inhibitors and activators (Monod et al. 1965; Koshland 1970; Kamoun et al. 1992; Fenton 2008; Liu and Nussinov 2016)

Allosteric effectors are biological in nature. They are inhibitors, often corresponding to products at the end of metabolic chains. They act by feedback on one of the first enzymes in the chain that produced them. They may also be small molecules that activate or inhibit (NAD^+; NADH; ATP; ADP; AMP), providing information on the cell's energy state.

Allosteric enzymes are made up of several sub-units (monomers), linked by numerous weak bonds. In addition to an active site, each sub-unit has one or two additional sites to which allosteric effectors are attached via weak bonds. These effectors are known as regulators.

1.4.1. *Allosteric enzyme kinetics*

In the case of these enzymes, Michaelis-Menten relationships are not found.

1.4.1.1. *K-type allosteric system*

Many of these catalysts produce sigmoid curves (representation v_i = f ([S]) – (Figure 1.7)). This is a feature of the cooperative action between the various monomers. In a K-type allosteric system, these enzymes are characterized by the fact that the first substrate molecules have difficulty binding to the enzyme. The higher the concentration, the greater the binding and transformation of the substrate: this is cooperativity or the positive homotropic effect of the substrate on its own binding.

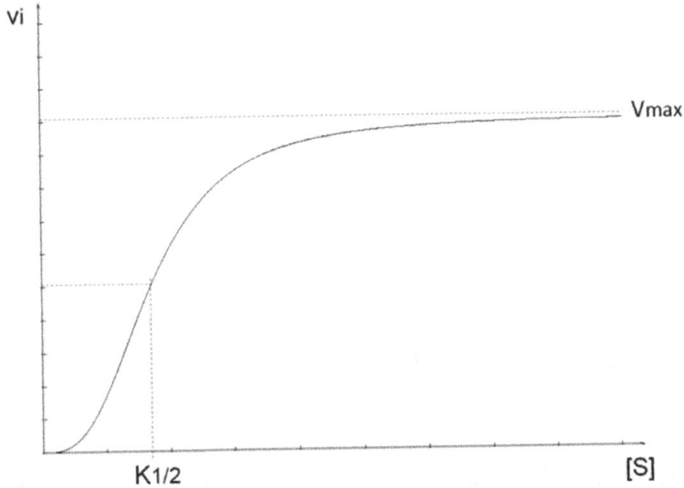

Figure 1.7. *Graph v_i = f ([S]) for a K-system allosteric enzyme (at least 10 different substrate concentrations are tested). A sigmoid curve is obtained for this allosteric kinetics. The $K_{1/2}$ designates the substrate concentration at which half the maximum velocity is reached*

The sigmoid equation is:

$$v_i = \frac{V_{max} \cdot [S]^h}{K_{1/2}{}^h + [S]^h}$$

with h being the Hill coefficient related to the degree of inflection of the curve.

When v_i is studied according to substrate concentration in the presence of an excess allosteric activator, an equilateral hyperbola is obtained. Its V_{max} does not change and the $K_{1/2}$ decreases compared with the curve obtained without an effector. This aspect is linked to the fact that the activating molecules will promote substrate binding to the enzyme and/or its transformation. Here again, this is a cooperative feature or positive heterotropic effect of the activator on substrate binding and/or its transformation.

When v_i is studied according to substrate concentration in the presence of an excess allosteric inhibitor, the curve obtained tends towards an equilateral hyperbola. Its V_{max} does not vary and the $K_{1/2}$ increases compared with the curve obtained without an effector. This aspect is linked to the fact that inhibitor molecules will hinder substrate binding to the enzyme and/or its transformation. Here again, this is a feature of cooperativity or a negative heterotropic effect of the inhibitor on substrate binding and/or transformation.

1.4.1.2. *V-type allosteric system*

For other catalysts, the curves (representation $v_i = f([S])$) have the appearance of an equilateral hyperbola. There is no indication of how the substrate is fixed at this stage and therefore of any cooperativity.

When v_i is studied according to substrate concentration in the presence of an excess allosteric activator, an equilateral hyperbola is obtained. Its V_{max} is increased and $K_{1/2}$ is unchanged compared with the curve without an effector. There is nothing to characterize the mode of binding of the activator and its effect on the substrate at this stage.

A study of v_i according to substrate concentration in the presence of an excess allosteric inhibitor results in an equilateral hyperbola with decreased V_{max} and no change in $K_{1/2}$, compared with the curve without an effector. There is nothing to characterize the mode of inhibitor binding and its effect on the substrate at this stage. It could be an absolute non-competitive inhibitor in a Michaelian system.

The curves (representation $v_i = f([S])$) for an enzyme operating in a V allosteric system without an effector, on the one hand, with a saturating concentration of activator, and with a saturating concentration of inhibitor, on the other hand, are shown in Figure 1.8.

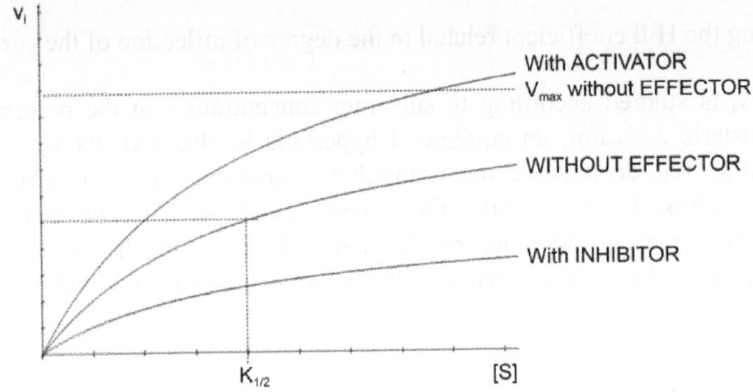

Figure 1.8. *Graph $v_i = f([S])$ for a V-type allosteric enzyme system: without effector; with saturating activator; with saturating inhibitor*

To characterize the cooperative effects generated by allosteric activators and inhibitors in a V system, further experiments are required. v_i must be represented as a function of effector concentration by setting a substrate concentration close to $K_{1/2}$. Sigmoids are obtained under these conditions and cooperativity traits can thus be observed. These experiments can also be carried out for effectors in a K system.

1.4.2. *Mechanism of allosteric enzyme regulation*

Two main models have been proposed to explain the cooperative interactions between sub-units of an allosteric enzyme. These are the symmetrical model of Monod, Wyman and Changeux, and the sequential model of Koshland.

1.4.2.1. *Symmetrical model*

In the Monod model, the enzyme exists in two forms. One form, R, has a high affinity for the substrate; the other, T, has a low affinity for the substrate. Both forms transform the substrate (with R >> T).

In the case of a K system, in the absence of substrate and any effector, the enzyme is predominantly in its T-form. The addition of substrate will produce a sigmoidal curve, progressively shifting the T/R equilibrium towards the R-form. Initially, in the presence of an activator (saturating concentration), the enzyme's T/R equilibrium will shift towards the R-form. The inhibitor binds preferentially to the T-form.

In a V system, there are initially as many T-forms as R-forms, and the substrate binds with as much affinity to the R-form as to the T-form, but is preferentially catalyzed by the R-form.

1.4.2.2. *Sequential model*

In the case of Koshland's concerted model, a tetramer, taken as an example, has four sub-units which can independently be in R or T-form. Transformation of one sub-unit to an R-form will facilitate that of subsequent sub-units. Here again, the substrate is better transformed by R (R >> T).

In short, depending on the allosteric case studied, the most appropriate model must be chosen.

The different modes of inhibition discussed are of vital importance in the development of new drugs. Their application to molecules used in therapeutics will be described in Chapters 2–4 of this book, with the respective advantages and disadvantages of the different modes of inhibition.

1.5. References

Fenton, A.W. (2008). Allostery: An illustrated definition for the 'second secret of life'. *Trends Biochem Sci*, 33(9), 420–425.

Hopkins, A.L. and Groom, C.R. (2002). The druggable genome. *Nat Rev Drug Discov*, 1(9), 727–730.

Kamoun, P., Leroux, J.-P., Demaugre, F. (1992). *Aide mémoire de biochimie*. Flammarion, Paris.

Kenneth, A.J. and Roger, S.G. (2011). The original Michaelis constant: Translation of the 1913 Michaelis-Menten paper. *Biochemistry*, 50(39), 8264–8269.

Koshland, D.E. (1970). The molecular basis for enzyme regulation. In *The Enzymes*, Boyer, P.D. (ed.). Academic Press, New York, 341–396.

Liu, J. and Nussinov, R. (2016). Allostery: An overview of its history, concepts, methods, and applications. *PLoS Comput Biol*, 12(6).

Michaelis, L. and Menten, M.L. (1913). Die kinetik der invertinwirkung. *Biochemische Zeitschrift*, 191349, 333–369.

Monod, J., Wyman, J., Changeux, J.P. (1965). On the nature of allosteric transitions: A plausible model. *J. Mol. Biol*, 88–118.

Rask-Andersen, M., Almén, M.S., Schiöth, H.B. (2011). Trends in the exploitation of novel drug targets. *Nat Rev Drug Discov.*, 10(8), 579–590.

Stein, R.L., Melandri, F., Dick, L. (1996). Kinetic characterization of the chymotryptic activity of the 20S proteasome. *Biochemistry*, 35(13), 3899–3908.

2

Targeted Viral and Microbial Enzymes

In the pharmaceutical field, there is an interest in working on enzymes implicated in infectious pathologies, genetic diseases or enzymes of the main metabolic pathways. A document drawn up in 2005 (Robertson 2005) provides a list of enzymes targeted in therapeutics. These chemotherapies are validated by the authorities controlling the marketing of drugs. This article listed 317 drugs approved by the FDA (Food and Drug Administration) that inhibit enzymes. More specifically, 71 enzymes were targeted: 48 human, 13 bacterial, five viral, four fungal and one protozoan. In 2007, an article (Imming et al. 2006) also listed 71 types of enzymes targeted in therapeutics, as well as four enzymes used as drugs directed against their specific substrate. The targeted enzymes differ slightly from one publication to another. In 2011, an article in *Nature Reviews* (Rask-Andersen et al. 2011) reported 124 enzymes targeted by drugs, including 97 soluble or membrane-bound human enzymes. This increase is linked to the development of kinase effectors and other enzymes involved in signaling pathways and in many cancers, the major pathologies of the 20th and 21st centuries.

Several websites dedicated to listing compounds in development or validated in therapeutics have even been created (Drug Bank; Therapeutic Drug Target Database TTD[1]).

Some enzymatic targets and associated inhibitors will be taken up again on the basis of the publications cited above and the works dedicated to this field (Kirkiacharian 2007, 2010), with more detailed data. It should be noted that not all the targets appear in these different documents. This shows how difficult it is to carry out this inventory work and the scarcity of such documents in databases. The ANSM website (Agence nationale de sécurité du médicament et des produits de

1 See: https://go.drugbank.com/.

santé), the EMA website and the FDA website were invaluable in compiling the following references.

The inhibition of an enzyme (E) by an inhibitor (I) can take different forms. The study of activity and the different modes of inhibition were presented in the first chapter of this book.

Enzyme inhibition may be reversible or irreversible.

In the first case, the bond (E-I) is weak (hydrogen, electrostatic, Van der Waals or hydrophobic bonds) and allows the enzyme to recover its activity after a more or less prolonged delay.

In the second case, it is irreversible, strong and covalent, and concerns the catalytic site of the enzyme, whose activity is regained only after a prolonged period, which is essential for the organism to develop a new biosynthesis. The choice of inhibitor will therefore depend on the pursued therapeutic objective.

A large number of enzymes are involved in human pathogenesis, as well as in parasites, bacteria, fungi and viruses. There is therefore an interest in having enzyme inhibitors that act selectively on a pathogenic species and have no action in the human species. In other cases, it is the patient's own enzymes that are the target of the drugs.

In Chapters 2 and 3, the enzymes involved in various pathologies are dealt with individually: the enzyme is mentioned, followed by its enzyme class (EC number for Enzyme Commission number in the international classification of enzymes).

The following list begins with the viral and microbial enzymes targeted in therapeutics.

2.1. Viral targets

2.1.1. *Herpes group viruses: DNA-dependent DNA polymerases (EC 2)*

These enzymes belong to class 2 in the enzyme classification. They are essential enzymes for the replication of viral genetic material and viral multiplication. Medicines directed more or less specifically against viral enzymes in the Herpes group (HSV, VZV and CMV) include *Acyclovir*, the best known, *Cidofovir*, *Famciclovir*, *Foscarnet*, *Ganciclovir*, *Valacyclovir* and *Valganciclovir*. Acyclovir (acycloguanosine) is a purine nucleoside analogue (guanine) that disrupts viral expression by competing with the nucleotide substrate of viral DNA polymerase. Acyclovir is a specific inhibitor of herpes viruses, with demonstrated in vitro activity against herpes simplex virus (HSV) types 1 and 2, varicella zoster virus (VZV), cytomegalovirus (CMV), Epstein–Barr virus (EBV) and human herpes virus type 6 (HHV-6). Acyclovir is only active in triphosphate form. Activating phosphorylation of the molecule is carried out by a specific viral enzyme: thymidine kinase (TK). This enzyme is only present in cells infected by the virus. Valacyclovir is the L-valine ester of Acyclovir. Activated, triphosphated Valganciclovir is a guanosine competitor. Cidofovir is a cytidine analogue. Foscarnet is a pyrophosphate analogue. It acts competitively by binding to the pyrophosphate binding site of the viral DNA polymerase. These molecules are shown in Figure 2.1.

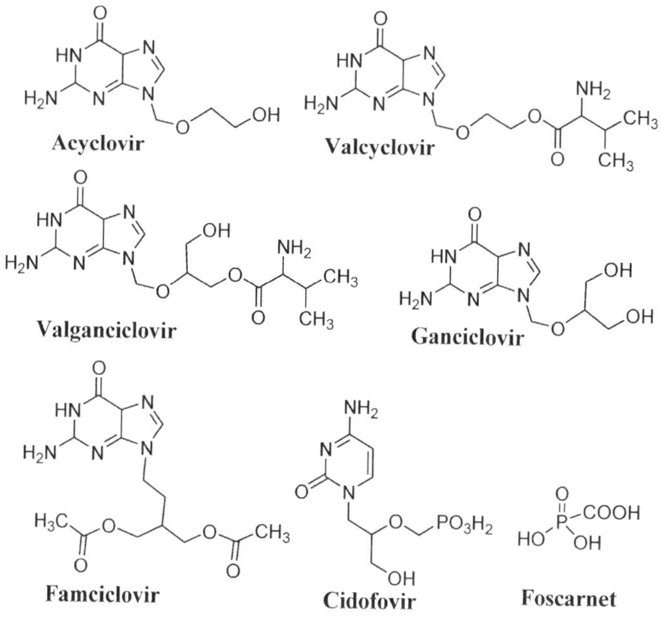

Figure 2.1. *Inhibitors of DNA-dependent DNA polymerases of Herpes group viruses*

2.1.2. *Influenza virus: exo-alpha-sialidase or neuraminidase (EC 3)*

This hydrolase enzyme is a glycosidase (class 3 in the enzyme classification). It is essential for separating virions from the host cell and disseminating the virus throughout the body. Two drugs are listed for their action against neuraminidase: *Oseltamivir* and *Zanamivir*.

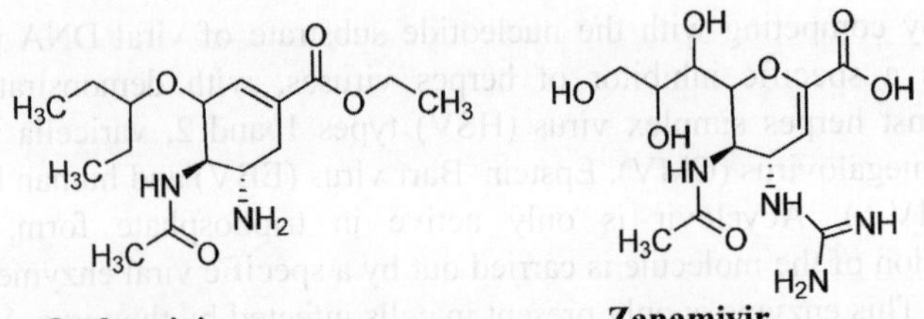

Oseltamivir **Zanamivir**

Neuraminidase inhibitors are not "new" drugs. They have been in the spotlight during recent flu epidemics, with fears of cross-species transmission. Zanamivir (Relenza®) was launched in 1999 and Oseltamivir (Tamiflu®) in 2002. They need to be administered at the right time, with good population coverage. However, these two drugs are often considered to have an unsatisfactory effect.

2.1.3. *HIV protease (EC 3)*

This enzyme belongs to the hydrolase family. It is involved in HIV multiplication. It is a member of the sub-subclass of aspartic acid proteases and has a different structure to human proteases. It consists of a homodimer of two 99 amino acid sub-units with C2-type symmetry. Each monomer contains the aspartic acid–threonine–glycine sequence (Asp-Thr-Gly), which constitutes the active symmetrical catalytic site. The catalytic triad of one monomer interacts through hydrogen bonds with the other monomer.

The N and C terminal ends contribute to the formation of an antiparallel four-stranded β-sheet structure, stabilizing the dimer by hydrogen bonding.

Protease cleaves non-functional polypeptides into proteins that are essential for viral maturation at the end of the multiplication cycle.

Mode of action: the protease's two residues, Asp25 and Asp25', react by activating a molecule of water, which then acts as a nucleophilic group, attacking the carbonyl function of the protein's peptide bond and releasing the peptide fragments (Figure 2.2).

Protease inhibitors (anti-proteases) act by attaching their hydroxyethylene or hydroxyethylamine group to the catalytic site of the HIV protease and inhibiting its activity. In this way, they block the formation of peptide fragments that are necessary for viral maturation and multiplication (formation of viruses with no infectious potential). This activity is seen in both recently or chronically infected cells (T4 (CD4+) lymphocytes and macrophages).

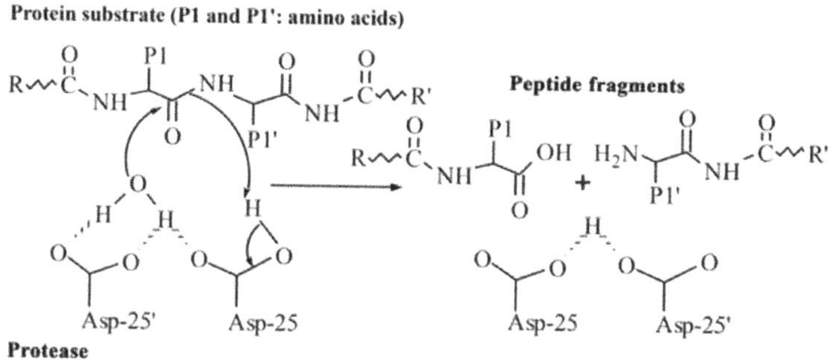

Figure 2.2. *Mode of action of HIV protease*

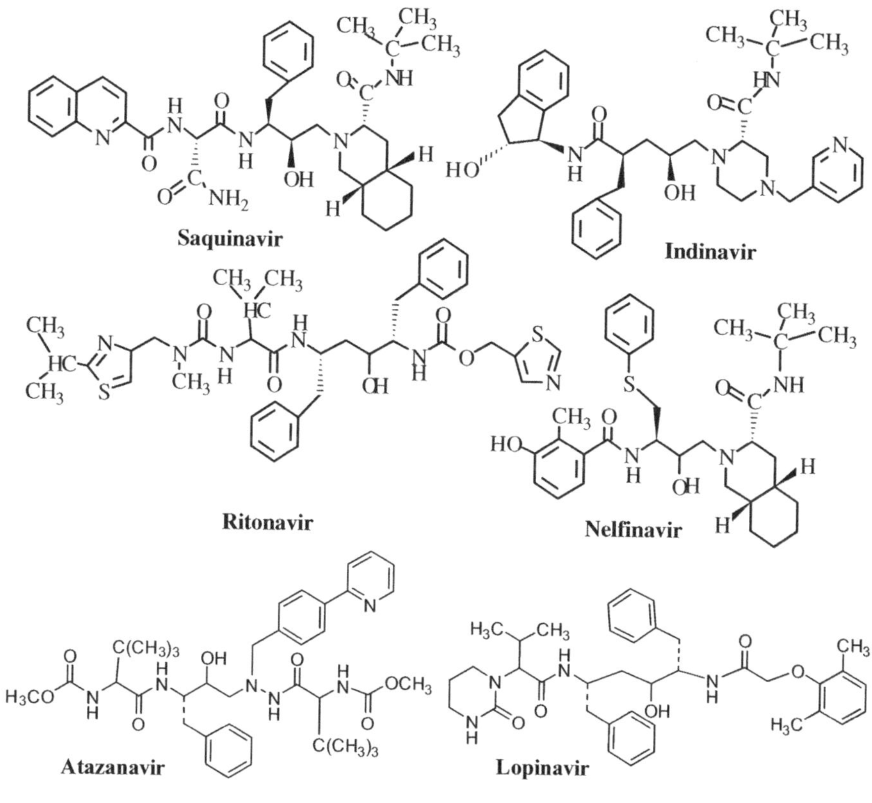

Figure 2.3. *Peptidomimetic HIV protease inhibitors*

Compounds that inhibit HIV protease fall into two groups: peptidomimetic inhibitors and non-peptidomimetic inhibitors.

Different peptidomimetic inhibitors are shown in Figure 2.3.

These peptide-like derivatives are characterized by poor intestinal absorption, low bioavailability, biological instability and rapid elimination.

Non-peptidomimetic inhibitors without a pseudo-peptide structure have been developed. Three representative options are currently on the market: Amprenavir, Darunavir and Tipranavir (Figure 2.4). These derivatives show little cross-resistance with peptidomimetic inhibitors.

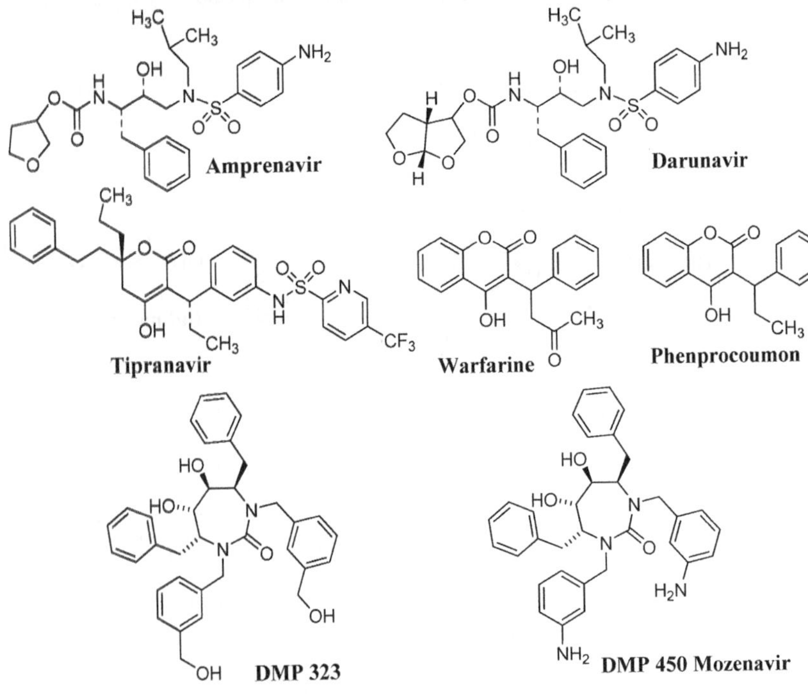

Figure 2.4. *Non-peptidomimetic HIV protease inhibitors*

Tipranavir is a 4-hydroxy-dihydropyrone derivative (Thaisrivongs et al. 1994). It was developed following the demonstration of HIV protease inhibitory activity by Warfarin (Bourinbaiar et al. 1993) and Phenprocoumon (Thaisrivongs et al. 1996). Various compounds substituted in position 3 of 4-hydroxycoumarin also have HIV protease inhibitory activity (Kirkiacharian et al. 2002). The structure–activity relationships of these derivatives (Aristoff 1998) are shown as follows:

– establishment of hydrogen bonds through their "enol" group in 4 at aspartate residues 25 and 25' of the enzyme's active site;

– occupation of pockets S1 and S1' and S2 of the enzyme by the two phenyl residues (Sn and Sn' designate the attachment pockets in Schechter and Berger's nomenclature at the enzyme level);

– formation of hydrogen bonds with the Ile 50 and Ile 50' residues, through the two oxygen atoms of the lactone function mimicking the role of a water molecule (Figure 2.5).

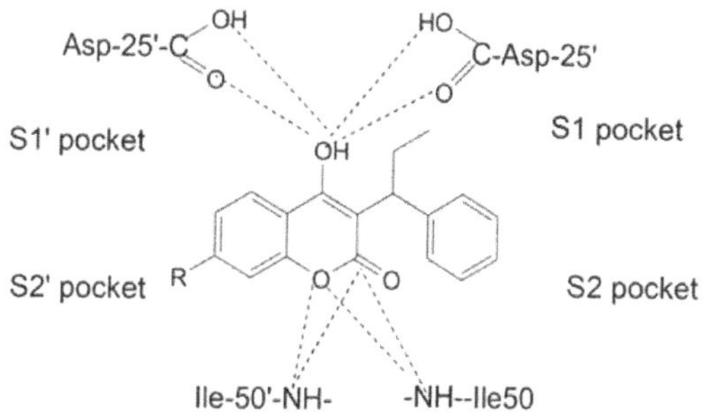

Figure 2.5. *Binding of a hydroxycoumarin to HIV protease*

This mode of action is also found in derivatives with a symmetrical cyclic urea structure, such as compound DMP-323 or DMP-450 (*Mozenavir*).

ASSESSMENT.– *There are nine drugs targeting PR: Atazanavir, Darunavir, Fosemprenavir, Indinavir, Lopinavir, Nelfinavir, Ritonavir, Saquinavir and Tipranavir. Fosamprenavir is a "prodrug" of Amprenavir. All these molecules target the active site of the enzyme and act as competitive inhibitors. They are selective and reversible inhibitors of the enzyme. All have been designed on the basis of the crystallographic structure of the enzyme and its mechanism of action. All these inhibitors are analogues of the transition state. For example, they have a non-hydrolysable isostere, such as a hydroxyethylene, hydroxyethyleneamine or hydroxyethylhydrazine isostere, which replaces the cleavable bond of the substrate. The hydroxy group interacts via hydrogen bonds with the catalytic aspartates. In all cases, the residues on either side of this type of isostere are optimized to improve the affinity of the PR for the inhibitors. Another important objective is to reduce the peptide character of the molecules. Anti-proteases are generally used in combination with nucleoside inhibitors of reverse transcriptase in bi and tritherapies, which are more effective.*

2.1.4. *HIV reverse transcriptase (RNA/DNA-dependent DNA polymerase – EC 2)*

Reverse transcriptase (RT) is one of the enzymes involved in the development of human HIV-1, leading to acquired immunodeficiency syndrome (AIDS). This disease, which is fatal if left untreated, has paved the way for the development of inhibitors of this enzyme's activity. It belongs to the family of transferases (nucleotide polymerases). It is a key enzyme in the viral reverse transcription. It produces double-stranded proviral DNA from single-stranded viral RNA, which can be integrated into the host's genetic material. This enzyme is characteristic of the Retrovirus family. Within the reverse transcription complex, it catalyzes the first stage of the virus' multiplication cycle in the host cell.

Two groups of drugs have been introduced into therapeutics: derivatives with a nucleoside structure and substances with a non-nucleoside structure.

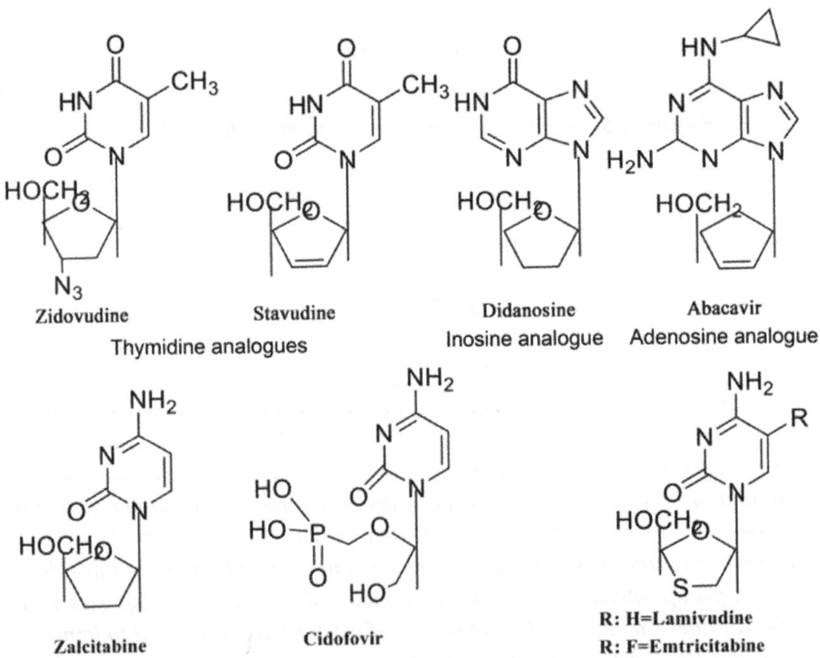

Figure 2.6. *Nucleoside analogues that inhibit HIV reverse transcriptase*

The first antiretroviral drugs were structural analogues of pyrimidine or purine nucleosides that had been modified in a number of ways. Their activity manifests itself after their intracellular transformation into nucleoside mono-, di- and triphosphates via kinases. These molecules inhibit viral reverse transcriptase and viral DNA synthesis. The chemical structure of antiviral nucleosides does not contain a 3'-hydroxyl group, preventing the formation of a bond with the 5' phosphate group of the following nucleotide. As a result, it is impossible to form the 3',5'-phosphodiester bond, thus interrupting the elongation of the viral DNA.

The structural modifications made to the pyrimidine and purine analogues of natural nucleosides involve the ribose molecule (Figure 2.6):

– replaced by a ring: a dihydrofuran (*Stavudine*), a tetrahydrofuran (*Didanosine*), a cyclopentene (*Abacavir*);

– replaced by a potential pentose ring (*Cidofovir* type).

Two other derivatives are purine nucleotide analogues: *Tenofovir disoproxil* (fumarate) and *Tenofovir alafenamide*. They contain acyclic ose derivatives (Figure 2.7).

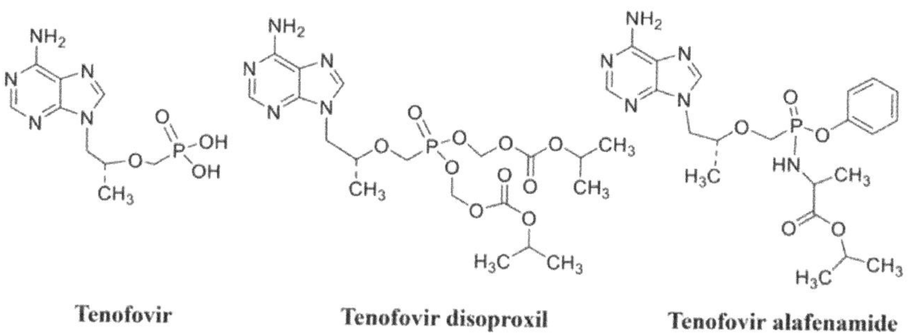

Figure 2.7. *Nucleotide analogues inhibiting HIV reverse transcriptase*

These precursors release the active fragment in the form of a nucleoside monophosphate, forming the triphosphate corresponding to the active metabolite, after two phosphorylation reactions. Their activity is used against HIV-1 viruses that are resistant to other antivirals.

Non-nucleoside derivatives constitute the second group of reverse transcriptase inhibitors. These are *Nevirapine*, *Delaviridine* and *Efavirenz*, as shown in Figure 2.8.

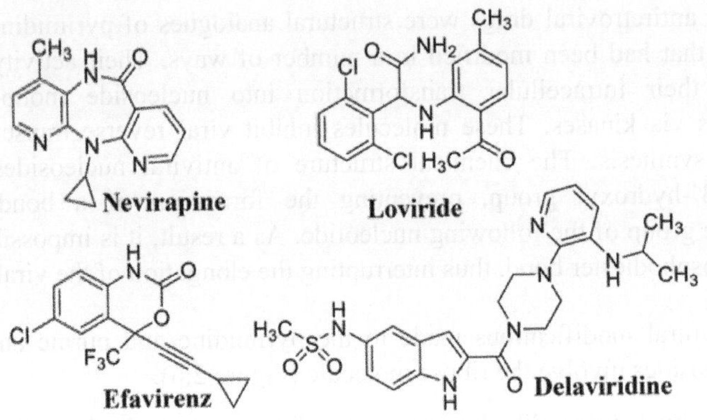

Figure 2.8. *Non-nucleoside HIV reverse transcriptase inhibitors*

These HIV-1-selective derivatives do not require phosphorylation by cellular enzymes to express their activity. They present little risk of cross-resistance with nucleoside analogues. However, in the case of certain viral mutations, resistance to the various agents in this group can occur.

ASSESSMENT.– *NRTI molecules act as false substrates. They become competitive inhibitors of RT after being converted into triphosphate molecules by cellular kinases at the hydroxyl level on the 5' carbon, for all eight nucleoside analogues. These compounds can compete with deoxyribonucleotides during their incorporation into the nascent viral DNA chain, but can also alter the binding of the substrate to RT. Phosphorylated derivatives are competitive analogues of nucleotide triphosphates that can disrupt viral DNA biosynthesis. They have all been designed with the 3' carbon of the deoxyribose sugar no longer carrying a hydroxyl function and are therefore unable to react with the phosphate of another nucleotide to create the phosphodiester bond, a key link in nucleic acid biosynthesis. Tenofovir, also a competitive inhibitor of RT, differs from its predecessors in that it has a phosphate group and no cyclic ose.*

NRTIs are positioned at the RT polymerase active site and prevent further elongation of the viral DNA.

NNRTIs act on a pocket in the p66 subunit of RT, located close to the active polymerase site. This binding leads to a conformational change that blocks the catalytic site in an inactive conformation, preventing DNA polymerase activity. Affiliation to this group of inhibitors is defined by the mode of action, not the chemical structure. The research carried out by De Clercq and Balzarini in 1992 on Nevirapine and other types of inhibitors like this did not allow any conclusions concerning the mechanism of action of these compounds on RT. RT is considered here as a Michaelian system. Depending on the polynucleotide substrates used, the kinetic results (Lineweaver and Burk method) vary. It can be said that these are non-competitive inhibitors that bind to a different site than the active site, as shown by the crystallographic studies carried out on RT/NNRTI complexes.

Pharmaceutical preparations include NRTI and NNRTI combinations such as Efavirenz + Emtricitabine + Tenofovir or Rilpivirine + Emtricitabine + Tenofovir. New HIV associations and treatments were also introduced but not reported here.

2.1.5. *HIV integrase (EC 2)*

This enzyme belongs to the transferase family. It is part of the viral reverse transcription complex.

The proviral DNA is integrated into the DNA of the host cell by binding the 5' end of the viral DNA to the 3' end of the host DNA. This step follows that of reverse transcription.

Three derivatives (Figure 2.9) are currently in use. They block the transfer of viral DNA into the host DNA. The three molecules are: *Elvitegravir* (quinolone carboxylic acid, structurally similar to antibacterials in this family) (Ramanathan et al. 2011; Lee et al. 2012), *Raltegravir* (substituted pyrimidinedione) and *Dolutegravir*. The latter integrase inhibitor is combined with two nucleosides, Abacavir and Lamivudine.

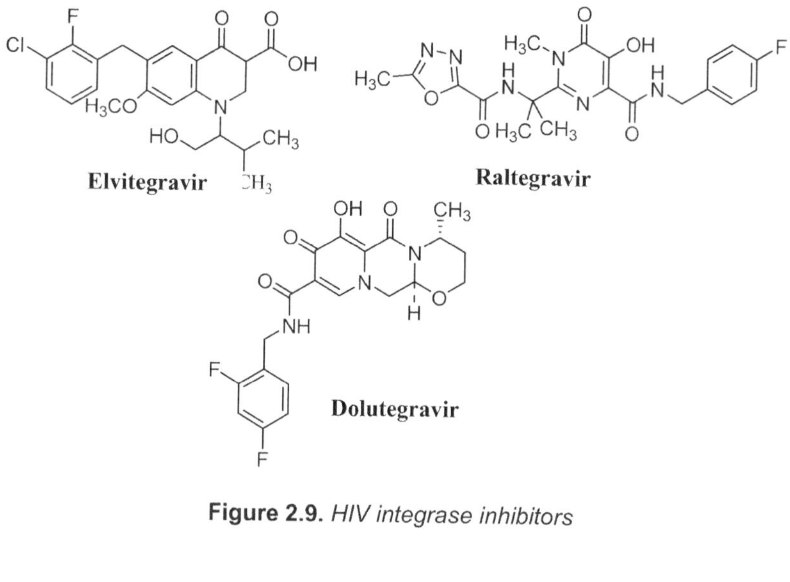

Figure 2.9. *HIV integrase inhibitors*

2.1.6. *Hepatitis C virus (HCV): RNA-dependent RNA polymerase (NS5B – EC 2) and viral protease (NS3-4A – EC 3)*

These are the targets of choice, since the polymerase is necessary for replication of the viral genome, a crucial stage in the formation of new virions, and the viral protease enables the maturation of viral polyproteins, as is the case for HIV protease. The former is an enzyme belonging to the class of enzymatic transferases (polymerases, EC 2), and the latter is a hydrolase (EC 3). Infection with the hepatitis C virus can lead to chronic infection, often leading to cirrhosis and hepatocellular carcinoma.

In terms of treatment, *Sofosbuvir* is a nucleotide polymerase inhibitor. *Dasabuvir* is a non-nucleoside inhibitor of this enzyme.

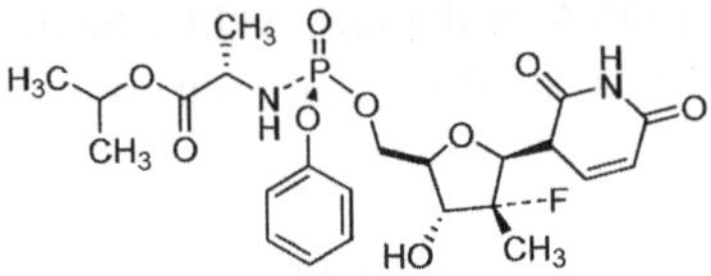

Sofosbuvir

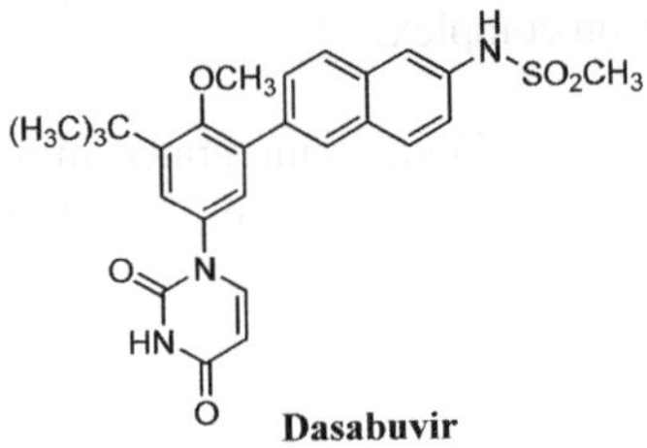

Dasabuvir

Protease inhibitors such as Boceprevir, Simeprevir and Telaprevir have not been marketed in France for several years.

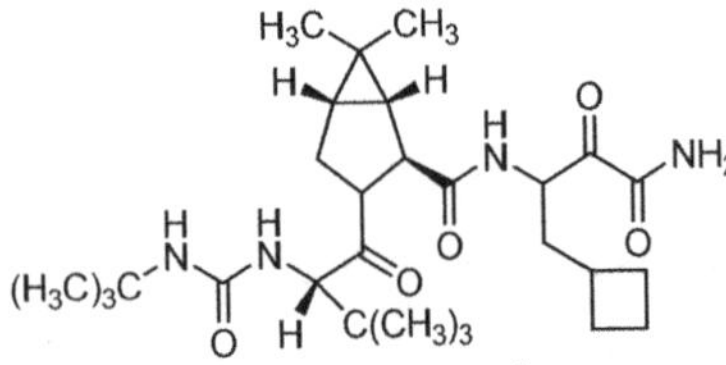

Boceprevir

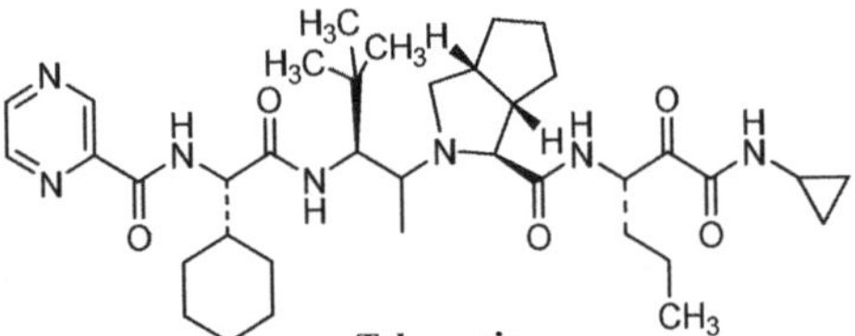

Telaprevir

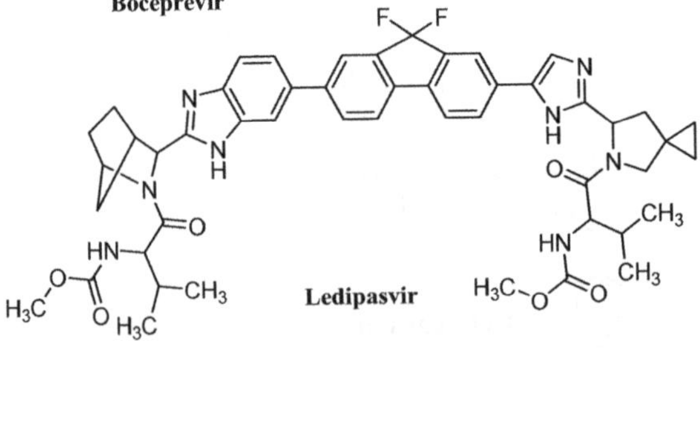

Ledipasvir

2.2. Bacterial targets

The sections devoted to the activity of antibiotics and their targets (spectrum of action) are given for information only. For much more selective and precise information on antibiotic therapy, readers can visit the website of the French National Agency for the Safety of Medicines (ANSM), NIH (National Institute of Health) or specialized pharmaceutical dictionaries.

2.2.1. *Specific target (action mainly focused on the* Mycobacterium tuberculosis *species)*

2.2.1.1. *Alanine racemase (EC 5)*

While the enzyme alanine racemase is present in many bacteria of the genus *Mycobacterium*, it is particularly targeted in attempts to halt the development of *Mycobacterium tuberculosis*. This bacterium is the cause of tuberculosis, which affects nine million people every year and causes more than a million deaths worldwide. This historically targeted enzyme is a homodimer that works with pyridoxal phosphate. It catalyzes the conversion of L-alanine to D-alanine, essential for the synthesis of bacterial peptidoglycan. *Cycloserine*, a cyclic analogue of alanine, is a substance produced by bacterial species of the genus *Streptomyces*.

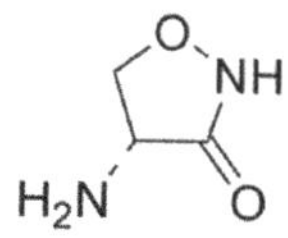

Cycloserine

This molecule has proved to be an effective antibiotic against alanine racemase. It binds covalently to the pyridoxal phosphate cofactor and prevents the enzyme from functioning properly. However, this product causes neurotoxicity, making it necessary to search for new compounds directed against this target.

2.2.1.2. *D-alanine-D-alanine ligase (EC 6)*

This enzyme is found in many bacteria. It is particularly significant to target in *Mycobacterium tuberculosis*. It is a dimer that catalyzes the assembly of two D-alanines using ATP to form a D-Ala-D-Ala dipeptide, essential for the formation of peptidoglycan. During this reaction, ADP and an inorganic phosphate are released. Cycloserine is an effective inhibitor of this enzyme, but the problems associated with the product's neurotoxicity, which has already mentioned, require further research. In particular, the ATP binding site is being explored (Triola et al. 2009).

2.2.1.3. *Enoyl (Acyl-Carrier-Protein = ACP) reductase (EC 1)*

This enzyme is still present in many bacteria. Several species possessing this protein are targeted (notably *Mycobacterium tuberculosis* and *Staphylococcus aureus*).

This enzyme catalyzes the NADH-dependent reduction of enoyl-ACP as part of the biosynthesis of fatty acids and mycolic acids.

Isoniazid is a highly effective inhibitor of the enzyme, which led to the development of Iproniazid and Nialamide as anti-tuberculosis drugs. The structure of these two molecules is shown in section 3.1.2.2 on MAOIs.

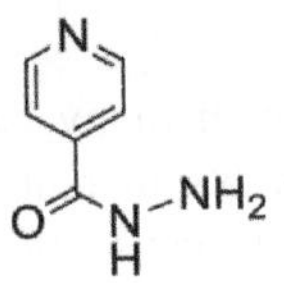

Isoniazid
(hydrazyde of isonicotinic acid)

Isoniazid is a "prodrug" activated by catalase, a peroxidase (EC 1). This drug needs to be monitored during treatment, as it can cause hepatitis.

Numerous molecules are currently being researched, including diazaborines, imidazole derivatives, naphthyridinones, pyridones and thiopyridines (Lu et al. 2012).

2.2.1.4. *Arabinosyltransferases (EC 2)*

In *Mycobacterium tuberculosis,* three arabinosyltransferase enzymes are involved in the synthesis of arabinogalactans and lipoarabinomannans. These molecules are essential to the bacterial structure and virulence of the microorganism. They are key structures in thc interaction between the pathogenic bacteria and the host.

Ethambutol is an antibacterial agent used in the treatment of tuberculosis. It is thought to act on three arabinosyltransferases: EmbA, EmbB and EmbC.

The presence of two chiral centers and a plane of symmetry leads to a racemic mixture that can be split into two enantiomers and an inherently inactive compound. The ethambutol used in therapy is the dextrorotatory derivative of 2,2'-ethylenediaminodibutanol or (S,S)-(+)-ethambutol. The other two stereoisomers are toxic.

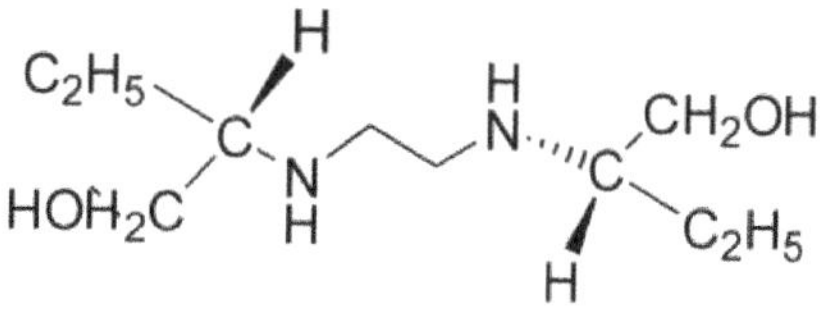

(S,S)-Ethambutol (antitubercular)

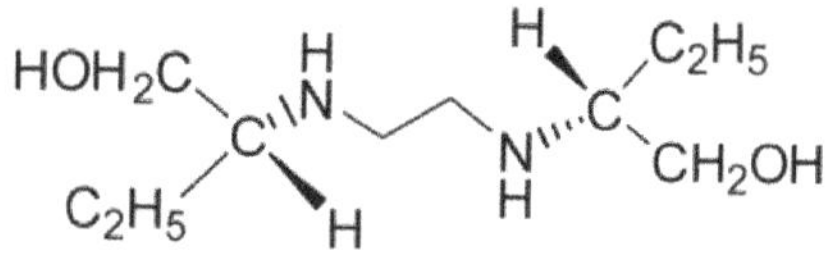

(R,R)-Ethambutol (toxic)

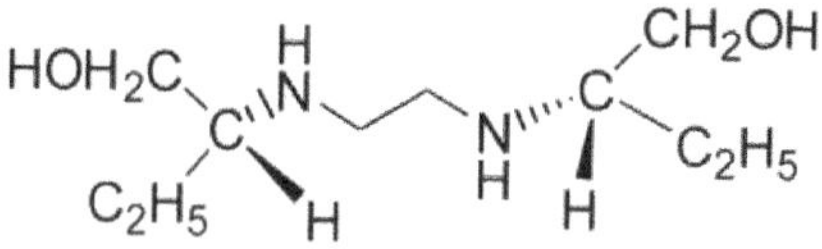

(R,S)-Ethambutol (toxic)

2.2.2. *General actions*

2.2.2.1. *Targeting DNA-dependent RNA polymerase (EC 2)*

These enzymes catalyze the initiation, elongation and termination of the synthesis of bacterial messenger RNA molecules from a transcribed strand of DNA. Prokaryotes have a single RNA polymerase with a simpler structure than those found in several isoforms in eukaryotes. As a result, these prokaryotic enzymes are good candidates for the development of antibiotics.

Inhibition of bacterial transcription of DNA to RNA by the polymerase is the mode of action of *Rifamycin B* derivatives developed by *Streptomyces mediterranei* (Floss and Yu 2005). They inhibit the enzyme by binding to one of its five sub-units. They have little effect on human RNA polymerase. Molecules in the rifamycin family are macrocyclic antibiotics (Figure 2.10). Hemisynthetic derivatives, and in particular *Rifampicin*, are used primarily for the treatment of tuberculosis. However, the emergence of resistant strains of *Mycobacterium tuberculosis*, as well as other mycobacteria, is making the need for new anti-tuberculosis drugs more and more pressing. Rifamycin is listed for the treatment of allergic conjunctivitis and bacterial keratitis, while Rifampicin is indicated for the treatment of brucellosis, atypical mycobacterial infections, leprosy, meningococcal meningitis, and above all tuberculosis, as we have already seen. Enterococci (aerobic GRAM +) are moderately sensitive to Rifamycin. GRAM – aerobes (enterobacteria, the majority of atypical mycobacteria and Pseudomonas) are not sensitive to this antibiotic.

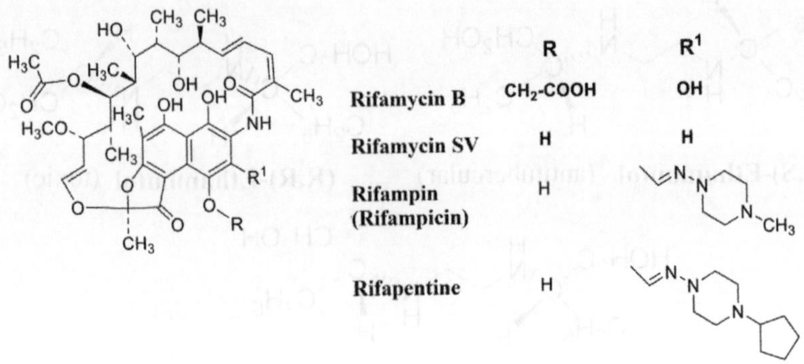

Figure 2.10. *Macrolides of the rifamycin family*

2.2.2.2. *Targeting DNA topoisomerase, gyrase (EC 5)*

Bacterial DNA-gyrase, or topoisomerase, is an enzyme involved in the supercoiling of bacterial DNA. The active form of the enzyme is a tetramer of the $\alpha_2\beta_2$ type, with the α sub-unit encoded by the GyrA gene and the β sub-unit encoded by the GyrB gene. Quinolone inhibitors have an antibacterial action. Quinolones act on this enzyme, which is involved in the replication and transcription phases of bacterial DNA (Figure 2.11). These bactericidal actions lead to germ lysis.

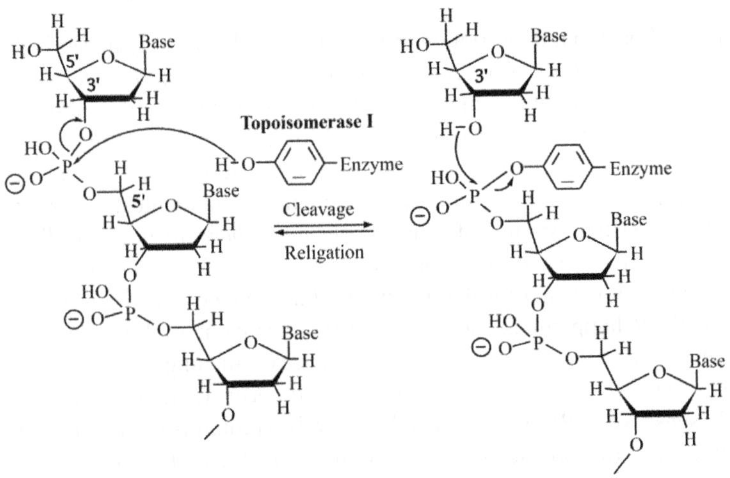

Figure 2.11. *Mechanism of topoisomerase I action*

Many quinolones (Figure 2.12), their isosteres, first-generation analogues and second-generation fluoroquinolones are currently used in therapy. The structures of some representatives are shown in Figure 2.13. Quinolones are active against aerobic GRAM – bacteria, including *Pseudomonas aeruginosa*, mycoplasmas and Rickettsia, and inactive against anaerobic GRAM – bacteria. Their action is moderate against GRAM + bacteria.

Numerous quinolones and their first-generation analogues (*nalidixic* acid, *pipemidic acid*) or second-generation fluoroquinolones (*Ofloxacin*, *Norfloxacin*, etc.) are currently used in therapy.

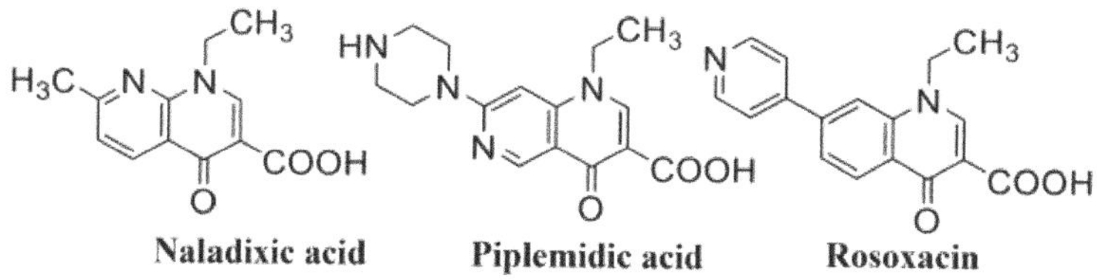

Figure 2.12. *First-generation quinolones*

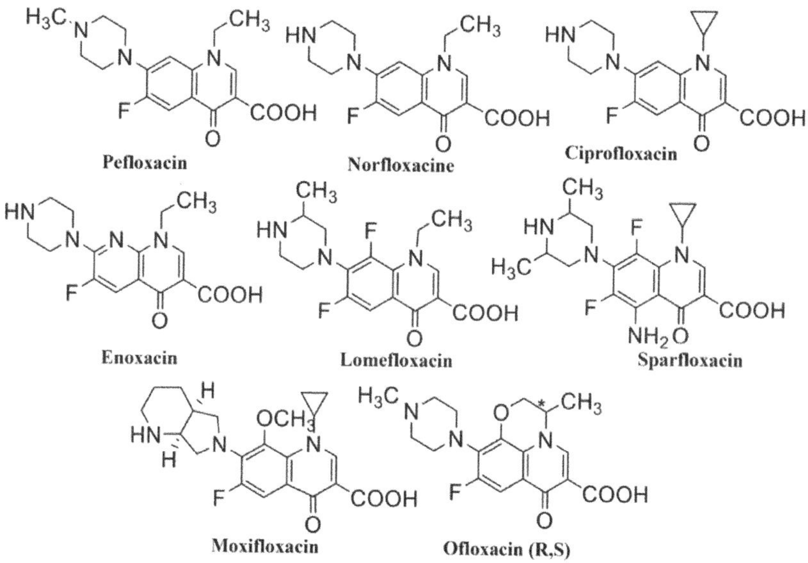

Figure 2.13. *Fluoroquinolones and second-generation analogues*

Quinolones can give rise to serious side effects (tendonitis, pain in the limbs, breathing difficulties, irritation, sudden and intense chest or back pain requiring emergency hospital treatment) at the start of treatment, or even later, after treatment has been stopped. Before using them, it is advisable to assess the benefit/risk ratio and consider substitution by an antibiotic with a similar antibacterial spectrum.

Other, more recent quinolones are also available, but have not been reported in this chapter.

2.2.2.3. *Targeting 30S or 50S ribosomal sub-units*

The two bacterial ribosomal sub-units are involved in the synthesis of proteins that are essential to the life of the microorganism, by carrying out the ordered polymerization of amino acids. The two bacterial sub-units, which differ from the two sub-units found in eukaryotes, are good candidates for therapeutic targeting.

2.2.2.3.1. Antibiotics acting on the 30S ribosomal sub-unit

Aminoglycosides (*Streptomycin, Neomycin, Kanamycin, Gentamycin*) and tetracyclines (Figure 2.14 for some tetracyclines) are two major families of antibiotics that disrupt the function of the 30S sub-unit, with varying degrees of activity and spectra of action.

Aminoglycosides mainly act on aerobic GRAM – bacteria and also on certain mycoplasmas and mycobacteria. They will be discussed in detail later.

Tetracyclines (Figure 2.14) are amphoteric substances that are poorly soluble in water (neutral pH). They have a broad spectrum of action on GRAM + and GRAM – bacteria. Their action on mycoplasma, Chlamydophila and Rickettsia is particularly noteworthy. It should be noted that resistance has appeared in many aerobic and anaerobic bacteria, and that these antibiotics can act on certain protozoa. Mycobacteria, Proteus and Pseudomonas are naturally resistant to tetracyclines. Other tetracyclines exist, including *Doxycycline, Metacycline, Rolitetracycline* and *Limecycline*, as well as new tetracyclines whose structures have not been reported.

Aminoglycosides are antibiotics with a heterosidic structure, consisting of a genin (aminocyclitol, diaminocyclohexane) linked to one or two specific amino acid molecules.

The genin corresponds to streptidine for *Streptomycin* and deoxystreptamine for *Kanamycins* and *Amikacin*. They have been isolated from microorganisms, some of which are mixtures (Kanamycins, Neomycins, Gentamycins). The first compound in this family, *Streptomycin* (Figure 2.15), isolated from *Streptomyces griseus* (actinomycetes), has an ose-ose-genin structure and its antibacterial activity mainly affects GRAM + and GRAM – germs, including *Mycobacterium tuberculosis*.

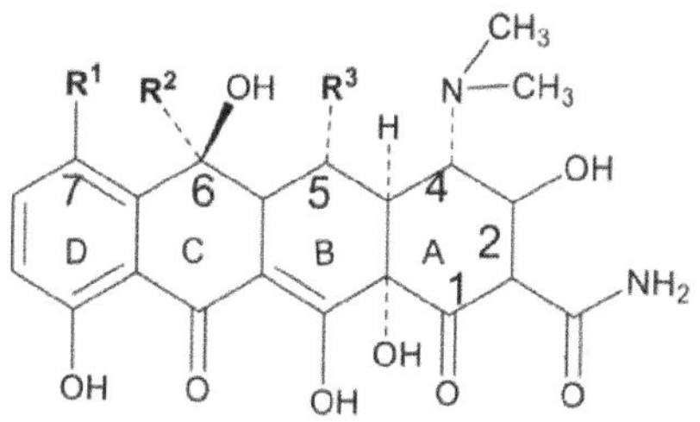

R^1	R^2	R^3	DCI
Cl	CH_3	H	Chlortetracycline
Cl	H	H	Demeclocycline
H	CH_3	H	Tetracycline
H	CH_3	OH	Oxytetracycline

Figure 2.14. *Structure of natural tetracyclines*

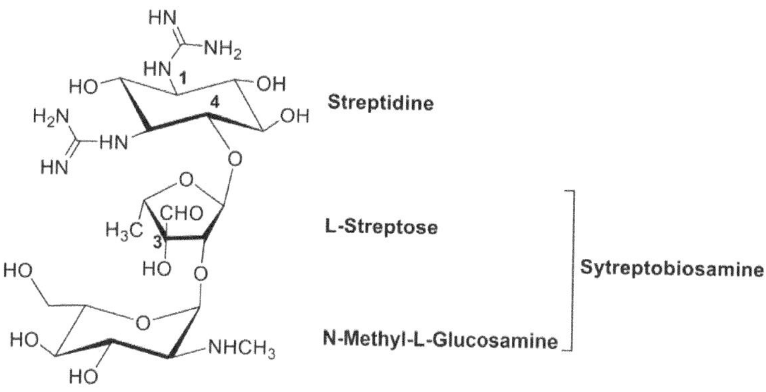

Figure 2.15. *Structure of Streptomycin*

The other naturally occurring aminoglycosides introduced into therapeutics were followed by hemisynthetic derivatives, following the loss of their antibiotic activity due to bacterial resistance. This loss of activity is the result of the modification of the functional groups involving the following enzymatic reactions:

– acetylation of amine groups by acetyltransferases;

– phosphorylation of hydroxy groups by phosphotransferases;

– adenylation of hydroxy groups by adenylate transferases.

Aminoglycosides have a bactericidal effect on multiplying germs. They disrupt bacterial protein synthesis by binding to the 30S units of ribosomes, resulting in the mistranslation of messenger RNA and the formation of proteins with an abnormal structure, leading to bacterial death.

The main representatives of this family, corresponding to two structural types of diosides, are the four and six diosides and the four and five diosides.

Four and six diosides

Kanamycins

Their structure type is ose-genin-ose.

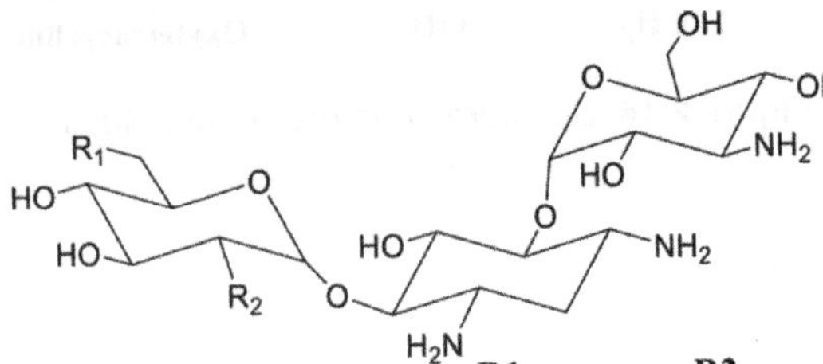

	R1	R2
Kanamycin A	**NH_2**	**OH**
Kanamycin B	**NH_2**	**NH_2**
Kanamycin C	**OH**	**NH_2**

These are replacement anti-tuberculosis drugs, with the antimicrobial spectrum mainly targeting *Mycobacterium tuberculosis* (see Gentamicin).

The main counter-indications are renal insufficiency, ototoxicity and nephrotoxicity.

Hemisynthetic derivatives of these compounds have been prepared to counteract bacterial resistance by modifying and/or protecting the functional groups.

Amikacin

The structure type is ose-genin-ose, corresponding to acylation of the amine function of Kanamycin A by 2-hydroxy-4-aminobutyric acid 2S.

The antibacterial spectrum targets *Staphylococcus* meti-S (methicillin-sensitive), GRAM – anaerobic bacilli (*Escherichia coli*, Proteus, Enterobacter, Serratia, Pseudomonas, Klebsiella, Acinetobacter). The indications are similar to those for Gentamycin.

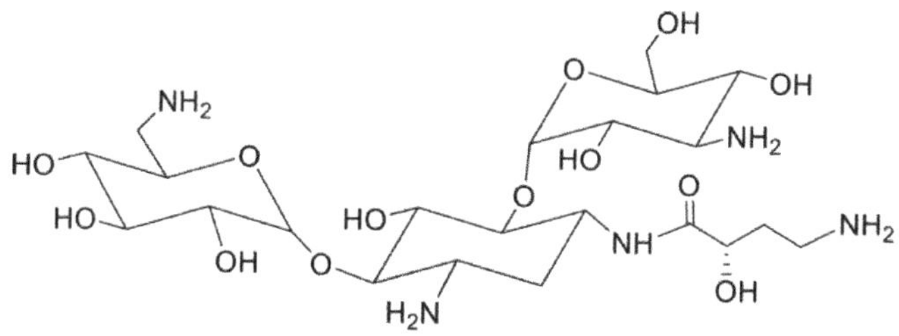

Amikacin

Dibekacin (sulfate)

Its structure corresponds to 3',4'-dideoxykanamycin B. It is active against enterobacteria, Pseudomonas and certain strains of staphylococci. This product is used to treat severe urinary tract infections.

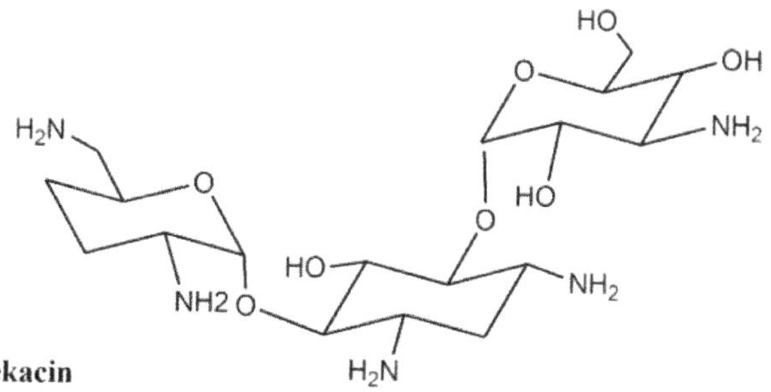

Dibekacin

Tobramycin (sulfate)

Its structure corresponds to 2'-amino 2',3'-dideoxykanamycin B. Its activity is similar to that of Gentamycin, but better on *Pseudomonas aeruginosa*.

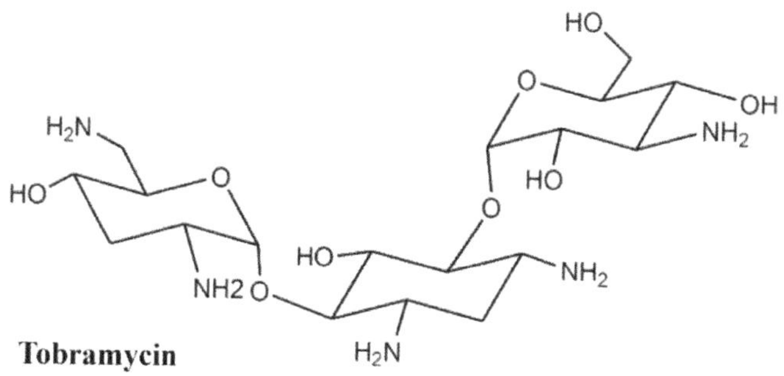

Tobramycin

Gentamycin (sulfate)

It is a mixture of three compounds with ose-genin-ose bonds, with gentosamine (4-methyl-3-methylaminodeoxypentose) and osamine (2,6-diamino-2,3,4,6-deoxyglucose) as oses.

The antibacterial spectrum is superior to that of other products in this group. This product is effective against GRAM – bacteria: *E. coli*, Enterobacter, Klebsiella, Serratia, Salmonella, Shigella and *Pseudomonas aeruginosa*, as well as GRAM + cocci: *Streptococcus* and *Staphylococcus* meti-S. This compound is used intramuscularly in urinary tract infections, pulmonary infections, severe meningitis and life-threatening septicemia.

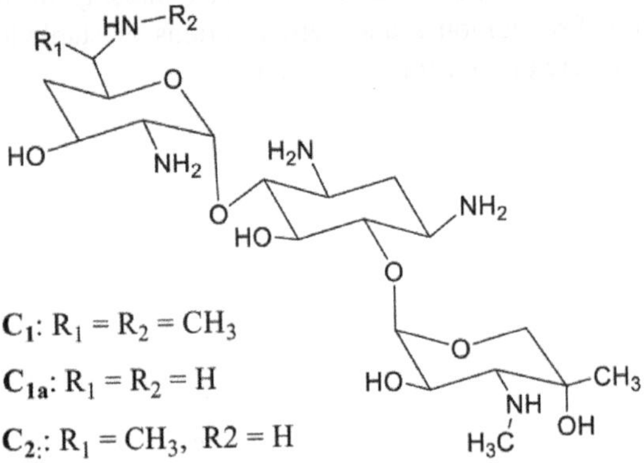

Gentamicin C_1: $R_1 = R_2 = CH_3$

Gentamicin C_{1a}: $R_1 = R_2 = H$

Gentamicin C_2:: $R_1 = CH_3$, $R2 = H$

Sisomicin (sulfate) and Netilmicin (sulfate)

Sisomicin (R = H) has a spectrum corresponding to that of Gentamycin, with superior activity in enterococcal infections. Netilmicin (R = C_2H_5) is the hemisynthetic derivative prepared by alkylation of the amine function of Sisomicin.

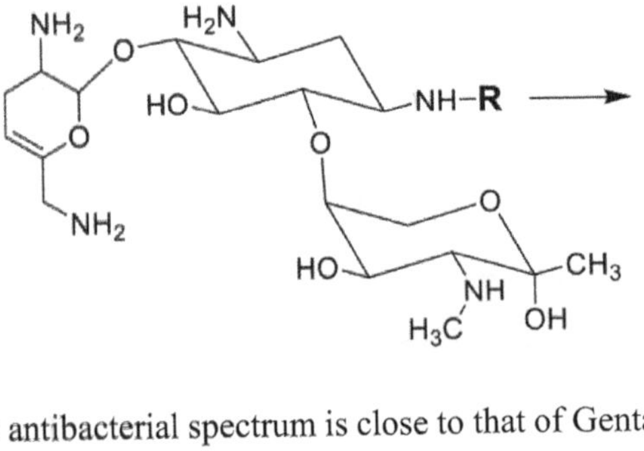

H: Sisomicin

C_2H_5: Netilmicin

Their antibacterial spectrum is close to that of Gentamycin.

NOTE.– The use of aminoglycosides is reserved for serious, life-threatening infections where other antibiotics cannot be used, and after careful assessment of the benefits against the risks of auditory and renal toxicity. They are used parenterally, as they are inactive orally due to their degradation. As they are water-soluble, they do not cross the blood–brain barrier and are inactive in the treatment of central nervous system infections.

Synergistic combinations of aminoglycosides with beta-lactams, whose effect on the wall favors the action of aminoglycosides, should be noted (in the case of streptococcal and staphylococcal endocarditis and GRAM – infections), as well as in immunocompromised patients with agranulocytosis.

Four and five diosides: neomycins, paromomycin (derivatives of ribostamycin, the latter not being used therapeutically)

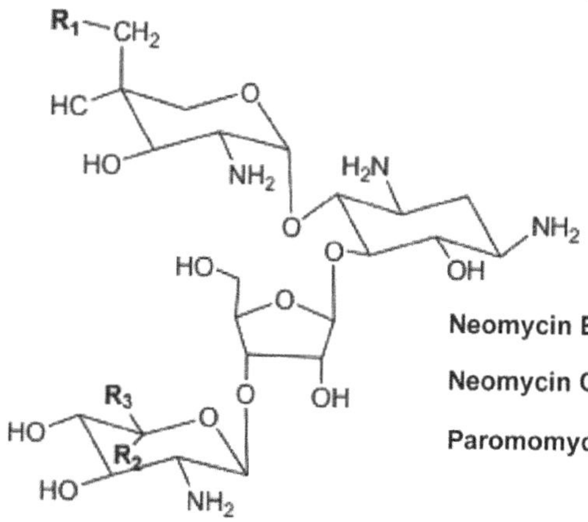

Neomycin B: $R_1 = NH_2$, $R_2 = CH_2NH_2$, $R3 = H$

Neomycin C: $R_1 = NH_2$, $R_2 = H_2$, $R_3 = CH_2NH_2$

Paromomycin: $R_1 = OH$, $R_2 = CH_2NH_2$, $R_3 = H$

These compounds are used in digestive disorders, infectious and parasitic diarrhea, as well as ear and eye infections.

2.2.2.3.2. Antibiotics acting on the 50S ribosomal sub-unit

Chloramphenicol, *Thiamphenicol* and *Florfenicol*, known as *phenicols*, are also inhibitors of protein synthesis.

By binding irreversibly to the 50S sub-unit, they prevent protein synthesis. Thiamphenicol is used in human medicine for salmonellosis, biliary infections, respiratory infections and *Haemophilus influenzae*. Enterococci, staphylococci, streptococci, Salmonella, Shigella, Aeromonas, Klebsiella, Enterobacter and *Escherichia coli* are sensitive to this bacteriostatic agent. Acinetobacter, Pseudomonas, Providencia, Serratia and mycobacteria are resistant to it. Chloramphenicol is used against the same pathogens, but its indications are limited to a small number of countries, due to hematological side effects. Florfenicol is a very broad-spectrum bacteriostatic used in veterinary medicine, especially in cattle and pigs.

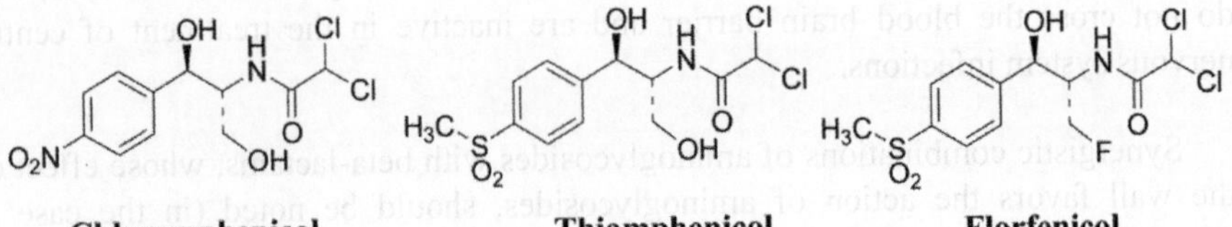

Chloramphenicol **Thiamphenicol** **Florfenicol**

Macrolides and *lincosamides* are structurally distinct molecules, whose modes of action lead them to be grouped together. They act as bacteriostatic agents, disrupting the function of the 50S sub-unit of bacterial ribosomes. These molecules are highly lipid-soluble and have very good tissue distribution.

Macrolides are macrocyclic derivatives with a lactone or olid residue. Macrolides, such as *Erythromycin*, *Azithromycin*, *Telithromycin*, *Spiramycin*, *Midecamycin*, *Josamycin* or *Tilmicosin* (the latter only used in veterinary medicine) are active against GRAM + bacteria (Corynebacterium, Listeria, Staphylococcus, Streptococcus, etc.) and GRAM – bacteria (Campylobacter, Actinobacillus, etc.). Anaerobic germs (Actinomyces, Clostridium, Bacteroides, etc.), germs with obligatory intracellular development, such as chlamydia, and certain mycoplasmas are also sensitive. Enterobacteriaceae and Pseudomonas are resistant.

Erythromycin was the first derivative of this family of antibiotics. Other compounds were subsequently introduced, including *Azithromycin*. *Clarithromycin*, *Dirithromycin* and *Cetromycin*, whose structures are not reported, are macrolides with the same mode of action as Erythromycin.

Erythronolide

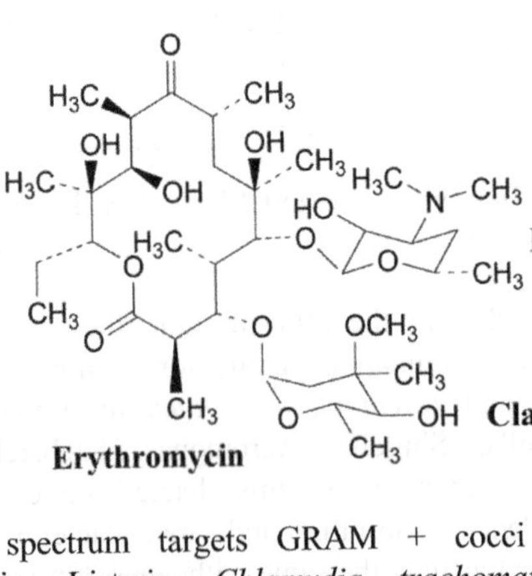

Mycaminosis (desosamine)

Cladinose

Erythromycin

The antibacterial spectrum targets GRAM + cocci and, in particular, streptococci. Clostridia, Listeria, *Chlamydia trachomatis*, Legionella and

Mycoplasma pneumoniae are also sensitive to this compound. Its indications are similar to penicillin.

Azithromycin no longer has the ketone function of Erythromycin and has an N-methyl residue.

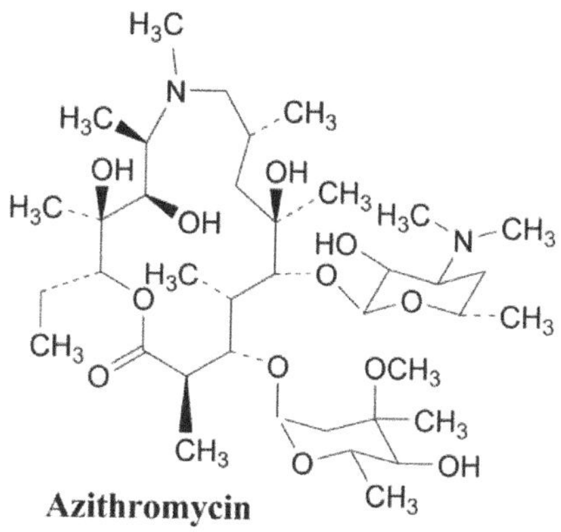

Azithromycin

This macrolide's spectrum extends to GRAM-negative bacteria: *Haemophilus influenzae, Branhamella catarrhalis* and intracellular germs: *Chlamydia pneumoniae, Mycoplasma pneumoniae.* Its main indications are sore throats, bronchitis and urethritis.

Telithromycin is a hemisynthetic derivative of erythromycin, belonging to a new group of compounds known as *Ketolides*.

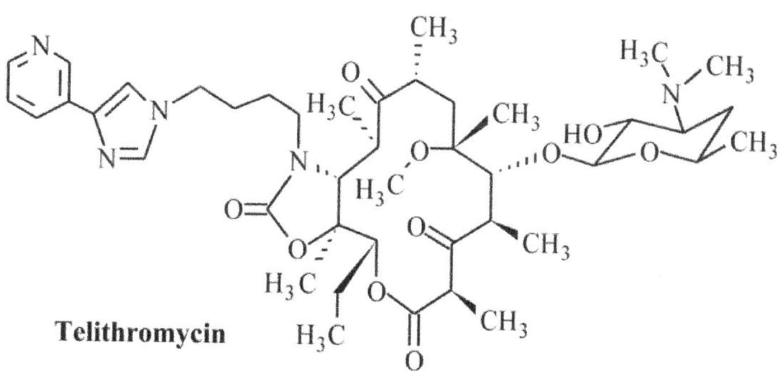

Telithromycin

It has antibacterial activity against GRAM-positive cocci (*Streptococcus pneumoniae*) and GRAM-negative bacteria (*Haemophilus influenzae, Moraxella catarrhalis*), as well as certain sensitive atypical germs. It is used to treat severe pneumonia or bronchitis, sinusitis and tonsillitis caused by sensitive germs resistant to penicillin or erythromycin.

Spiramycin is a mixture of three derivatives (macrolides), isolated from *Streptomyces ambofaciens*. Its mode of action is the inhibition of protein synthesis through its action on the 50S ribosome. It acts on GRAM-positive aerobic germs (indications for macrolides) and is used in the prophylaxis of meningococcal meningitis.

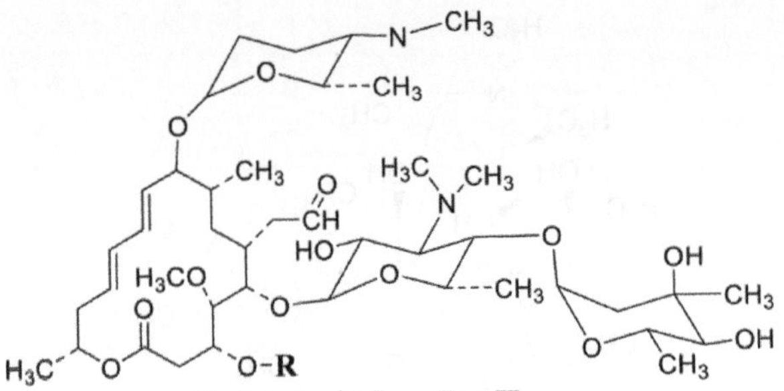

Spiramycin I **R = H**
Spiramycin II **R = $COCH_3$**
Spiramycin III **R = $COCH_2CH_3$**

Midecamycin, which is similar to the previous antibiotic, has the general indications of macrolides for sensitive germs.

Josamycin is also a macrolide. The compound is extracted from *Streptomyces narbonensis* var. *josamyceticus*. Its spectrum of activity and therapeutic indications are similar to those of Erythromycin.

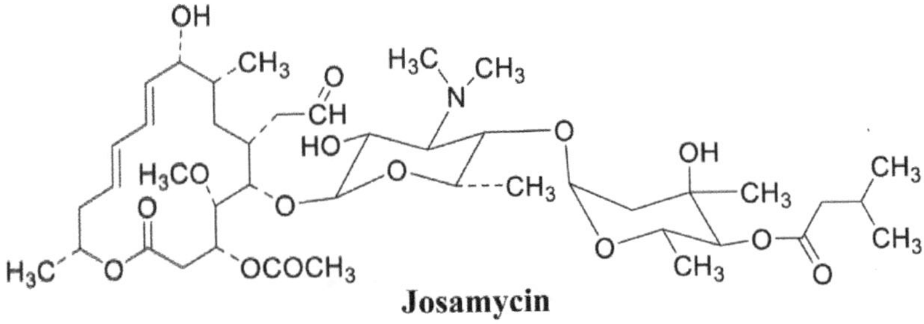

Josamycin

Lincosamides (Figure 2.16), including *Lincomycin* and *Clindamycin*, are active against GRAM + bacteria, anaerobes and mycoplasmas. Many GRAM – bacteria are resistant. Lincosamides can antagonize the action of Chloramphenicol, Erythromycin and Tilmicosin.

The lincosamides are related to the macrolides and have two representatives. Their structure is derived from N-methytlproline and galactopyranose thiomethyl ether.

Lincomycin is an antibiotic isolated from *Streptomyces lincolnensis*. Its spectrum includes GRAM + cocci, GRAM + bacilli and, in particular, staphylococci. Its mechanism of action is bactericidal, acting like Erythromycin on protein synthesis at the 50S sub-unit of the ribosome. Its indications are GRAM + cocci infections (streptococci and staphylococci) and the treatment of osteomyelitis (due to very good bone diffusion).

Clindamycin is a hemisynthetic derivative. It has a broader spectrum, encompassing Cocci GRAM – (gonococcus) and bacteroids. Its indications are similar to those of Lincomycin.

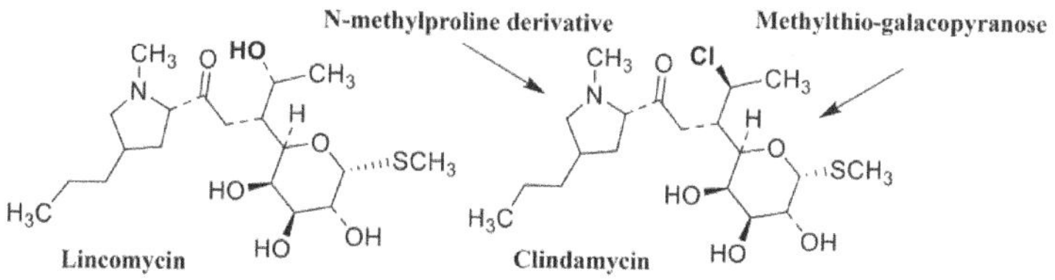

Figure 2.16. *Structure of lincosamides*

2.2.2.4. *Carboxypeptidases (EC 3), transpeptidases (EC 2), transglycosylases (EC 2), endopeptidases (EC 3) (Penicilline Binding Proteins – PBPs) and beta-lactamases (EC 3)*

2.2.2.4.1. PBPs

The latter are involved in peptidoglycan synthesis. They belong to the Penicillin-Binding Proteins (PBP) family. They catalyze transglycosylation, transpeptidation, carboxypeptidation and endopeptidation reactions (Sauvage et al. 2008). Certain molecules from the penicillin family are able to bind to these enzymes once they have crossed the bacterial wall.

Molecules with a β-lactam ring have a structural analogy with the D-Ala-D-Ala unit of peptidoglycan. They are able to bind to the enzymatic domains recognizing this dipeptide part.

The enzyme opens the beta-lactam ring, but is unable to complete the reaction. A stable acyl-enzyme complex is formed. This is why β-lactams are considered to be suicidal substrates for these enzymes. Antimicrobial activity results from a covalent bond between the antibiotic and the active sites of one or more groups of enzymes corresponding to PBPs (Duguide 1946; Park and Strominger 1957; Spratt 1983; Tipper 1985).

Antibiotics containing the β-lactam group (penicillins, cephalosporins and monobactams) are PBP inhibitors. They exert their antibacterial activity by inhibiting the peptidoglycan assembly stages of their walls (cross-linked polymers). When they are well structured, these polymers impart bacteria rigidity and resistance (Barreteau et al. 2008; Drawz and Bonomo 2010).

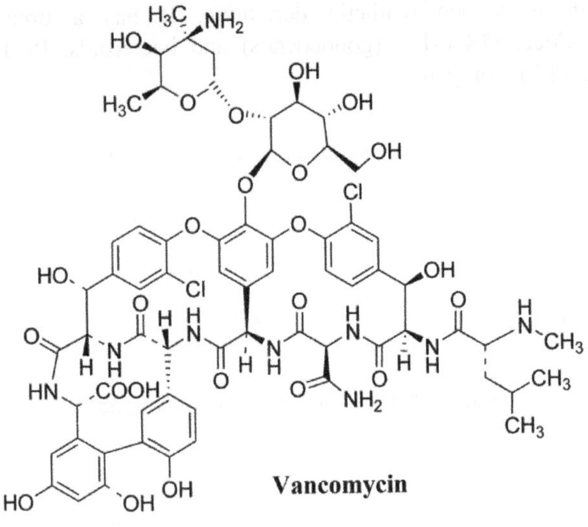

Vancomycin

Penicillins, such as Benzylpenicillin and its derivatives, are sensitive to beta-lactamases which, when present, inactivate these antibiotics. Aminopenicillins (such as *Ampicillin* and *Amoxicillin*) and cephalosporins (such as *Cefalexin, Ceftiofur, Cefotaxime*) are generally more resistant to bacterial beta-lactamases. It should be noted that resistance to penicillins exists in certain GRAM + bacteria, in particular certain strains of *Staphylococcus aureus*. These bacteria produce extracellular beta-lactamases. Many GRAM – germs are also resistant. Benzylpenicillin derivatives are a priori active against a number of GRAM + bacteria (including beta-hemolytic streptococci), certain GRAM + and GRAM – anaerobic bacteria (Clostridium, Fusobacterium or Bacteroides) and, finally, GRAM – germs (Haemophilus, Pasteurella, Actinobacillus).

Vancomycin is an amphoteric glycopeptide. This antibiotic is extracted from *Streptomyces orientalis* cultures. Its structure comprises a heptapeptide, amino acid residues, aromatic rings and a disaccharide consisting of glucose and vancosamine. Its antibiotic activity is exerted through inhibition of bacterial wall biosynthesis, by binding to the terminal D-Ala-D-Ala group, with inhibition of transglycosylases and transpeptidases.

2.2.2.4.2. Beta-lactamases

These are peptidases that split the lactam bond, transforming penicillins and cephalosporins into derivatives with no antibacterial effect.

Bacterial resistance to treatment of infections with penicillins and cephalosporins is linked to the secretion of beta-lactamases and cephalosporinases by bacteria. These serine enzymes act on the beta-lactam ring of these antibiotics and lead to their opening, forming derivatives with no antibacterial activity (Tipper 1985; Sauvage et al. 2008; Drawz 2010) (Figure 2.17).

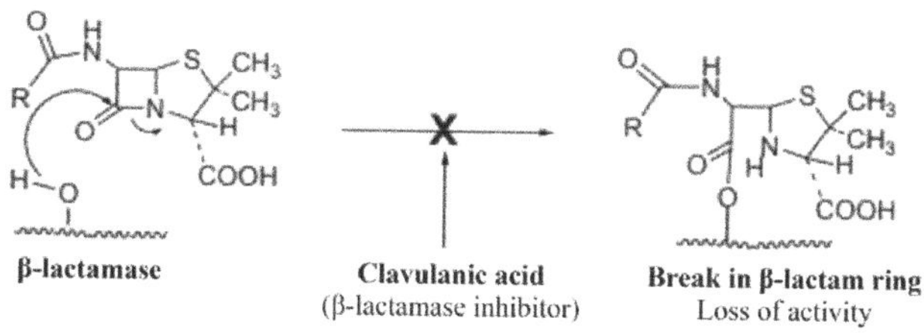

Figure 2.17. *Action of a beta-lactamase inhibitor*

The combination of beta-lactamase inhibitors with penicillins and cephalosporins prevents the degradation of the beta-lactam ring of these antibiotics and enables antibacterial activity to be maintained.

Solutions to the problem of bacterial resistance through the secretion of beta-lactamases have been approached in two ways, as shown in Figure 2.18:

– preparation of a large number of new beta-lactam penicillins and cephalosporins, represented in a condensed form by their typical general structure in Figure 2.18;

– synthesis of beta-lactamase inhibitors that prevent the opening of the β-lactam ring and consequently the loss of activity.

Beta-lactam antibiotics

Antibiotics containing a beta-lactam group are derivatives of the four-membered nitrogen heterocycle or azetidine, forming azetidine-2-one, which leads to most penicillins, cephalosporins and beta-lactamase inhibitors (Figure 2.18).

Structural considerations

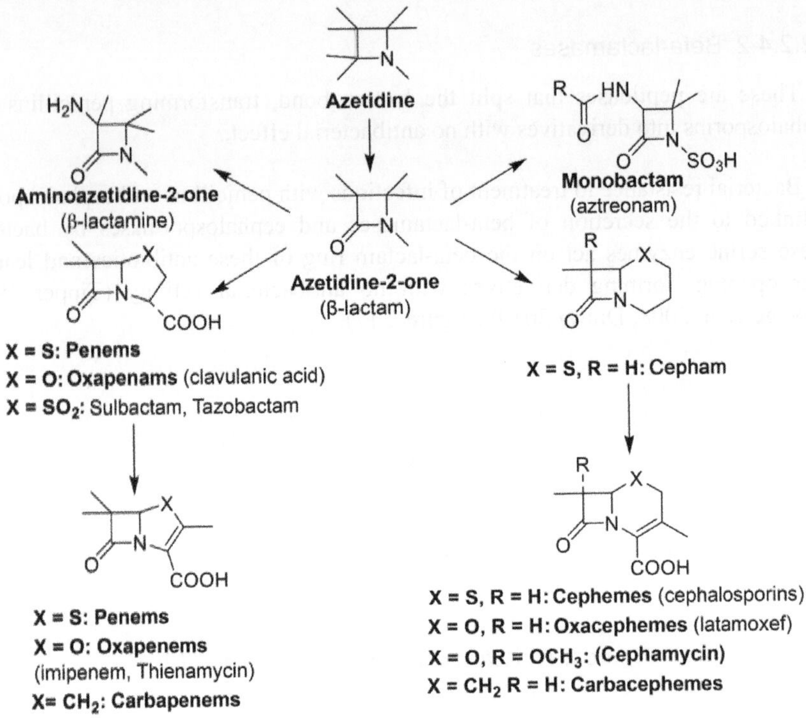

Figure 2.18. *General structure of penicillins, cephalosporins and beta-lactamase inhibitors*

Penicillins

The structure of penicillins comprises an azetidine-2-one ring (beta-lactam group) and a thiazolidine ring with a common side. They are amides of 6-aminopenicillanic acid with three chiral centers. Two natural derivatives, *Benzylpenicillin* and *Phenoxymethylpenicillin* (penicillin V), are obtained by extraction from the culture media of various penicilliums. The same applies to 6-amino-penicillanic acid. The latter is also prepared by hydrolysis of the amide bonds of various natural penicillins and by total synthesis. It is used as a raw material in the preparation of various synthetic penicillins by amidification of the 6-amine group.

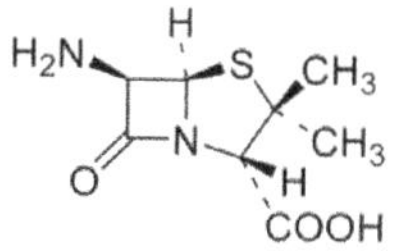

9-aminopenicillanic acid

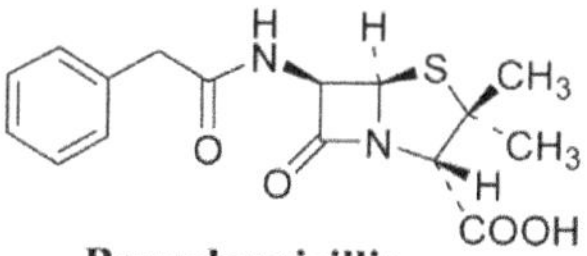

Benzylpenicillin

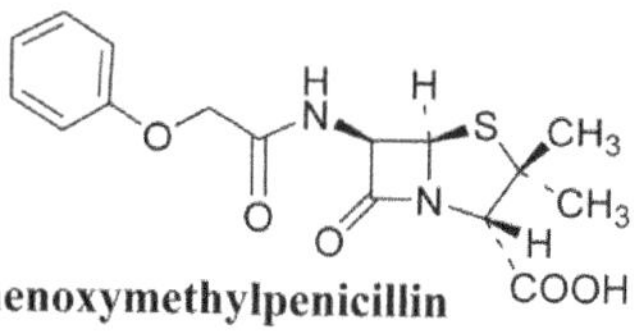

Phenoxymethylpenicillin

Benzylpenicillin is used parenterally in the form of soluble sodium or potassium salts, due to its instability in the acidic environment of the stomach. The acyl group amplifying the electronegativity of the oxygen in the beta-lactam's carbonyl group facilitates its degradation in an acid environment, as shown in the diagram below. Phenoxymethylpenicillin, on the contrary, is orally active due to its electron-withdrawing radical, which reduces this electronegativity and stabilizes the molecule in an acidic environment.

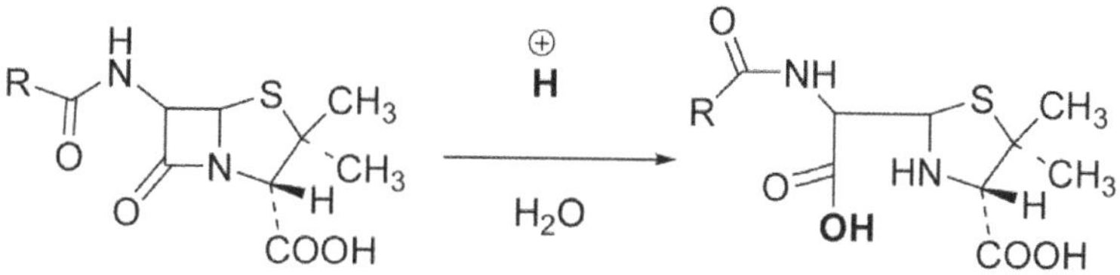

Penicillin

The activity of these derivatives is bacteriostatic and is exerted by inhibition of transpeptidase on young germs during the development of peptidoglycan, the main constituent of the bacterial wall. This is formed by long heteropolysaccharide chains linked together by peptide chains.

The antibacterial spectrum targets GRAM + cocci: streptococci, pneumococci and non-penicillinase-producing staphylococci. Among GRAM – cocci, only *Neisseria gonorrhoeae* (gonococcus) and meningitidis (meningococcus) are sensitive, due to the structure of their walls. They are also active against GRAM + bacilli: *Bacillus anthracis*, *Corynebacterium diphtheriae*, Listeria; certain spirochetes and leptospires: Borrelia, Leptospira, Treponema; as well as GRAM + anaerobes: Actinomyces, Clostridia (with the exception of *Clostridium difficile*) and Fusobacterium. They are inactive against penicillinase-producing staphylococci, enterobacteria and tuberculosis bacilli.

The rapid elimination of these derivatives has led to research into the preparation of amine benzylpenicillinates that are poorly soluble in water and eliminated more slowly for intramuscular use: procaine benzylpenicillinate, benethamine benzylpenicillinate (eliminated in 24 hours) and *N,N'*-dibenzyl-ethylenediamine benzylpenicillinate (eliminated in two weeks).

Their indications are identical to those of penicillin G, but they are of interest in the treatment of syphilis and rheumatic fever. Despite the slight improvement in initial treatments brought about by these modifications, the resistance of penicillinase-secreting germs renders these derivatives inactive. Research has since focused on the preparation of *hemisynthetic penicillins*, which meet the growing needs of anti-infective therapy and are made possible from 6-aminopenicillanic acid.

The objectives were as follows:

– prepare compounds resistant to the action of germ penicillinases;

– have access to broad-spectrum compounds;

– access to orally active derivatives.

Penicillinase-resistant penicillins or group "M" penicillins, Methicillin and Isoxazolylpenicillin

Penicillinase-resistant penicillins, or group "M" penicillins, Methicillin and Isoxazolylpenicillin can be used orally and parenterally.

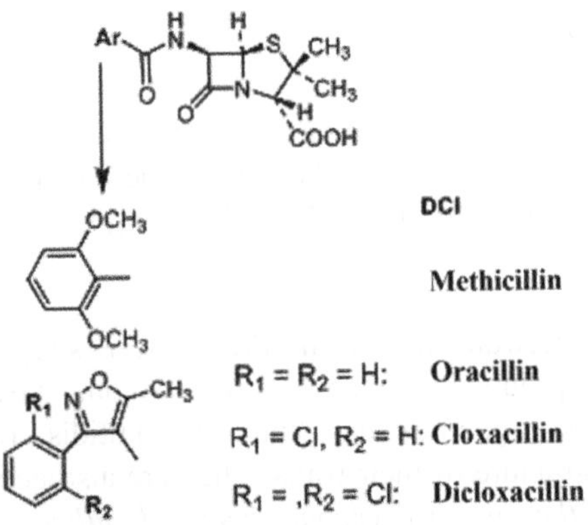

These compounds resist the inhibitory action of lactamases, due to steric hindrance exerted by the substituents protecting the lactam group. However, only Oxacillin and Cloxacillin are still used. The other compounds are rarely used, due to the acquired resistance of certain germs, known as methicillin-resistant (or meti-R)

germs. They are used in the case of methicillin-sensitive (or meti-S) staphylococcal infections and in combination, in serious cases, with aminoglycosides and Fosfomycin.

Extended-spectrum penicillins or group "A" penicillins

Extended-spectrum penicillins, or group "A" penicillins, correspond to the *Acylaminopenicillins: Ampicillin* and its derivatives. These compounds are Proampicillins, prepared by blocking the amine function or by esterification of the carboxyl group in position 2. By attaching a substituted aromatic residue that attracts electrons, Ampicillin is stable in an acidic environment and can be used orally (absorption of around 40%). The pharmacokinetic characteristics of Ampicillin are improved by the attachment of a hydroxy group, resulting in *Amoxicillin*, which is stable in acidic media and widely used orally.

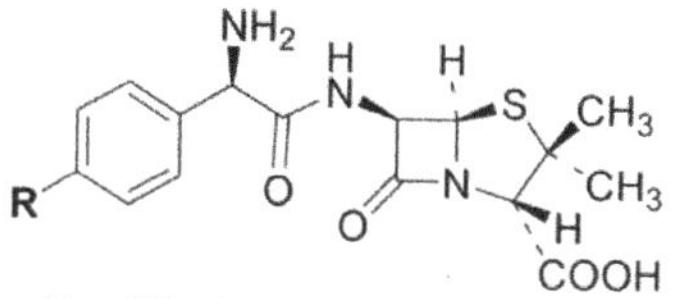

R = H: Ampicillin
R = OH: Amoxicillin

The spectrum of these derivatives corresponds to that of benzylpenicillin, but extended to include certain GRAM – bacilli: *Escherichia coli*, *Proteus mirabilis*, Salmonella, Shigella, *Haemophilus influenzae*, *Bordetella pertussis*, Brucella and *Vibrio cholerae*. However, more and more strains of *Escherichia coli*, *Proteus*, Salmonella, Shigella and *Hemophilus influenzae* have become resistant through the production of penicillinases. The same is true for gonococci, staphylococci and pneumococci.

Combining Amoxicillin with clavulanic acid (a beta-lactamase inhibitor) improves antibacterial activity.

Carboxypenicillins, Carbenicillin and Ticarcillin

Carboxypenicillins, *Carbenicillin* and *Ticarcillin* result from the replacement of the amine group of Ampicillin by a carboxylic residue, used in the form of water-soluble sodium salts. Their spectrum is broader than that of amino-penicillins, extending to GRAM-positive germs: Acinetobacter, Pseudomonas, Enterobacter, indole-positive Proteus, Colibacillus, Citrobacter and Haemophilus. These derivatives are used parenterally in infections caused by aminopenicillin-resistant Pseudomonas and Proteus and in systemic and bone infections caused by sensitive anaerobic germs.

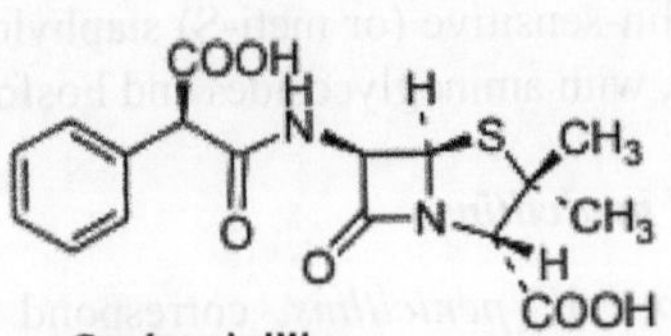

Carbenicillin

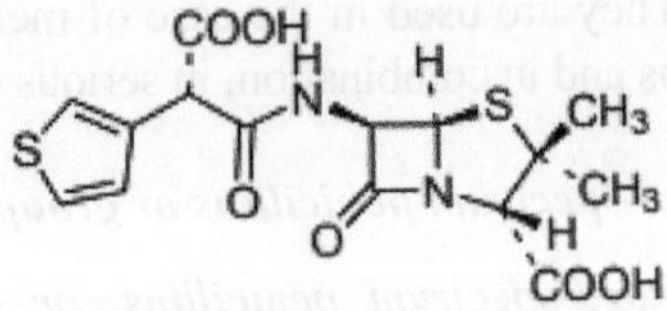

Ticarcillin

Combining Ticarcillin with clavulanic acid improves antibacterial activity.

Ureidopenicillins

Ureidopenicillins are obtained by substitution of the amine group of Ampicillin. They are effective against GRAM – bacilli, particularly *Pseudomonas aeruginosa.* They are used in infections caused by germs resistant to carboxypenicillins.

Cephalosporins

Cephalosporins were isolated from cultures of *Cephalosporium acremonium.* The strain produces three types of antibiotic, including cephalosporin C, corresponding to the amide of δ-aminoadipic acid and 7-aminocephalosporanic acid. The structure comprises an azetidine 2-one ring and a 1,3-dihydrothiazine ring with a common side.

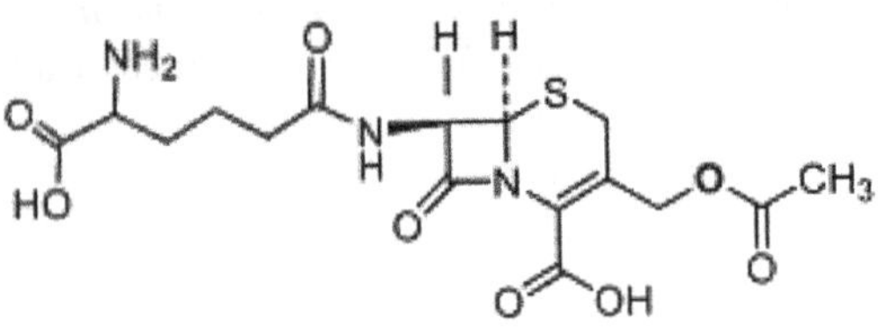

Cephalosporin C

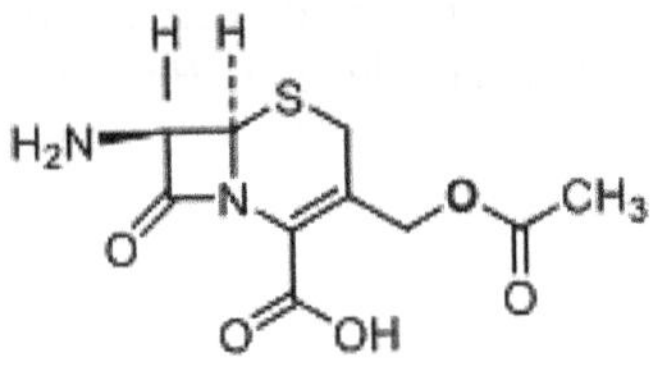

7-aminocephalosporanic acid

The chemical properties of cephalosporins are similar to those of penicillin G. However, the beta-lactam ring of cephalosporins is more resistant to the action of penicillinase. The carboxyl group, which is more acidic, yields salts and leads to the formation of a lactone with the hydroxymethyl group attached to carbon 3 after hydrolysis. The acetoxymethyl group at 3, corresponding to the acetate of an unsaturated alcohol (allyl type), is very reactive and, under the influence of a tertiary amine and the departure of the acetate group, an ammonium salt is formed, which is salified by the intramolecular carboxyl with the formation of a betaine structure (zwitterion in the case of certain hemisynthesis cephalosporins). The R1 radical in the following formula is identical to the chemical structure of the starting cephalosporin.

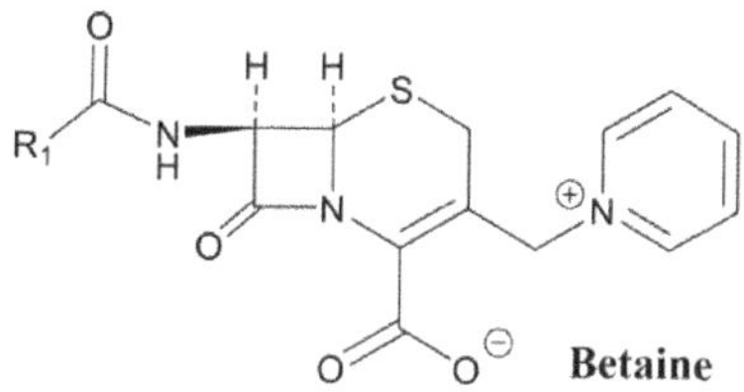

Betaine

Cephalosporin C is weakly active, and a large number of hemisynthetic derivatives have been prepared from 7-aminocephalosporanic acid. They are classified into generations according to their antibacterial spectrum. A few representatives will be given here due to their large number.

Hemisynthetic cephalosporins

The research was performed with the following objectives:

– extend the bacterial spectrum to enterobacteria and to get lower MICs (minimum inhibition concentration);

– increase stability against beta-lactamases;

– improve pharmacokinetic characteristics;

– access to orally active derivatives.

These results were made possible by the methods developed to obtain 7-aminocephalosporanic acid. The derivatives prepared are shown in Figure 2.19. Esterification of the carboxyl function of cephalosporins increases their lipophilicity and gives them good intestinal absorption, with the possibility of oral administration.

Classification of cephalosporins

Cephalosporins are classified into "generations" according to the following criteria:

– breadth of bacterial spectrum;

– stability against beta-lactamases;

– intrinsic antibacterial activity expressed by MIC, particularly against enterobacteria;

– cephalosporins are less active against staphylococci than penicillins in the Methicillin–Oxacillin group.

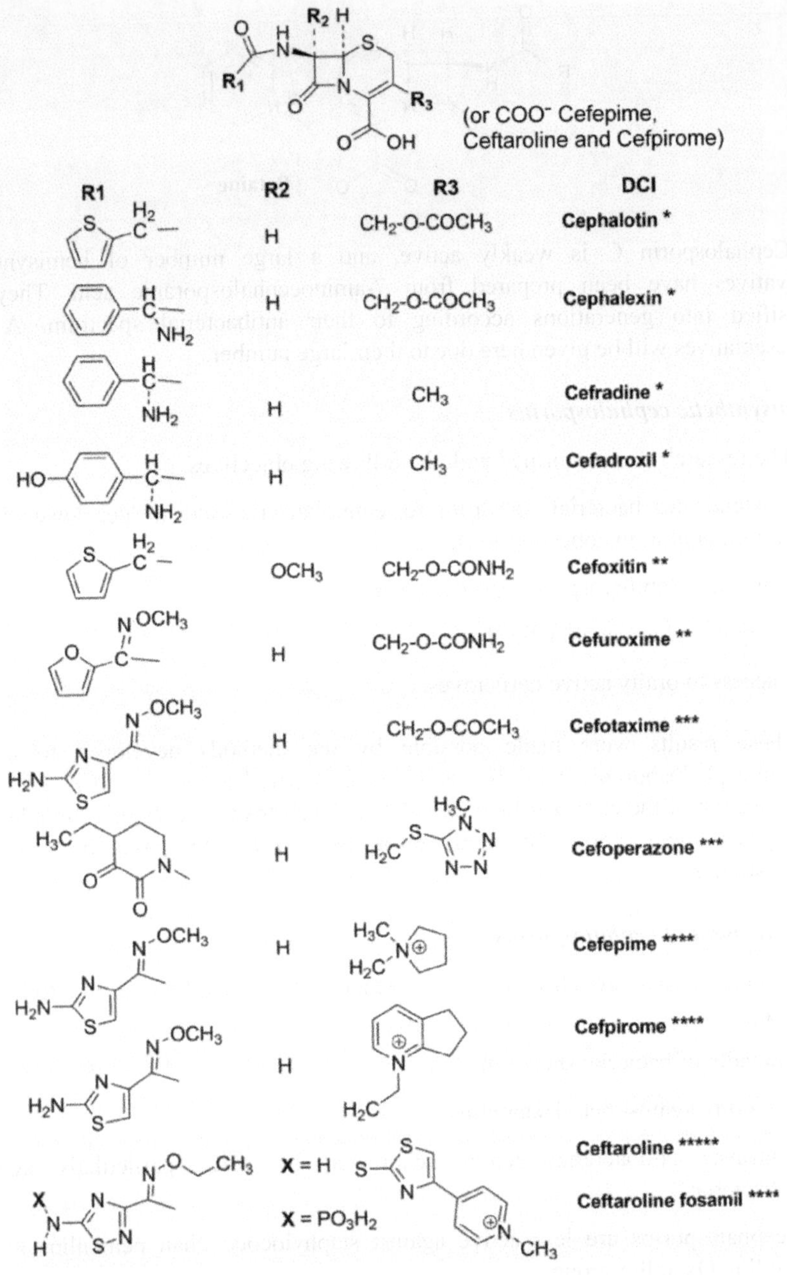

Figure 2.19. *General structure of cephalosporins. The number of stars represents the generation to which it belongs. Cefepime and Ceftaroline are zwitterions with ionization of the carboxyl group*

Antibacterial spectrum of first-generation cephalosporins (C1G)

First-generation cephalosporins have:

– increased resistance to penicillinase in staphylococci;

– activity against a number of GRAM-negative bacilli among non-beta-lactamase-producing Enterobacteriaceae;

– anti-pyocyanic activity (*Pseudomonas aeruginosa*);

– weaker activity than penicillin G against streptococci and pneumococcus.

Antibacterial spectrum of second-generation cephalosporins (C2G)

These derivatives have:

– better activity on sensitive strains than C1G;

– greater resistance to cephalosporinases.

Antibacterial spectrum of third-generation cephalosporins (C3G)

These products accentuate the advantages of the previous generations:

– better activity on susceptible strains, with lower MICs than C1G and C2G;

– greater resistance to inactivation by cephalosporinases;

– very good activity against Enterobacteriaceae, including beta-lactamase secretors;

– against staphylococcus, they are no better than the old cephalosporins and are less active than penicillins in the Meticillin–Oxacillin group;

– *Haemophilus*-producing beta-lactamase is constantly susceptible;

– in hospitals, there are strains that produce extended-spectrum beta-lactamases (ESBL), which hydrolyze most beta-lactam antibiotics of this generation.

Antibacterial spectrum of fourth-generation cephalosporins (C4G)

Among the C4Gs, *Cefpirome* and *Cefepime* have a “zwitterion” structure, giving them:

– better penetration through the porins in the bacterial wall. The massive influx of antibiotics increases the likelihood of them reaching their target;

– the very broad spectrum compared with previous derivatives also gives them activity against the majority of GRAM + anaerobic bacteria and enterobacteria;

– cefepime: stable against beta-lactamases, active against meti-S staphylococci, streptococci and most low MIC Enterobacteriaceae (*Escherichia coli*, Salmonella, Shigella, *Proteus mirabilis* and *vulgaris*, Providencia, Serratia). Less active on *Pseudomonas aeruginosa*;

– cefpirome: same activity as above.

These two derivatives are used to treat respiratory infections, pyelonephritis, septicemia, pneumonia, urinary tract infections and biliary tract infections. They are reserved for the treatment of hospital infections.

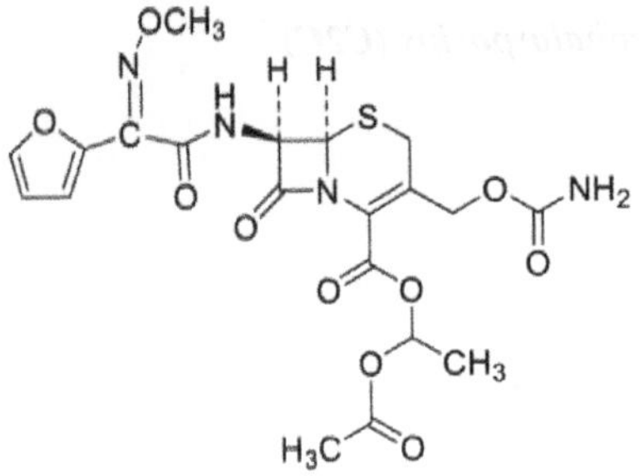

Cefuroxime axetil **

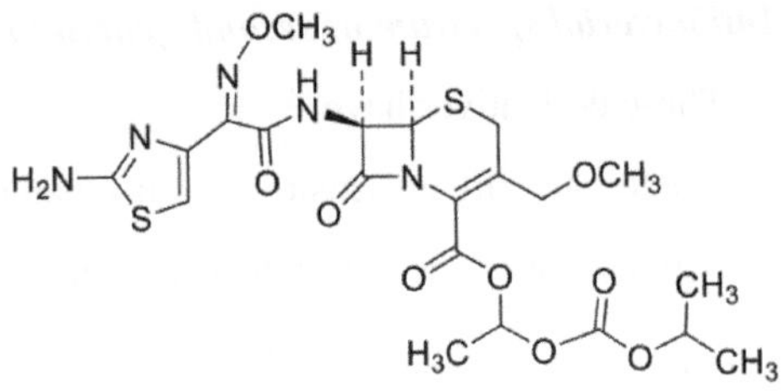

Cefpodoxime proxetile ***

Antibacterial spectrum of fifth-generation cephalosporins (C5G)

Unlike most previous cephalosporins, *Ceftaroline* is active against methicillin-resistant staphylococci. *Cefuroxime fosamil* N-phosphate derivative is the water-soluble precursor and is reserved for parenteral administration. This compound is also active against a large number of GRAM-positive and GRAM-negative pathogenic bacteria. This compound is listed for complicated skin and soft tissue infections, as well as community-acquired pneumonia. In vitro data show that the following species are not susceptible to this antibiotic: *Chlamydophila* spp., *Legionella* spp., *Mycoplasma* spp., *Proteus* spp. and *Pseudomonas aeruginosa*.

NOTE.– Although not all cephalosporins can be used orally, esterification of the carboxylic acid function increases lipophilicity and ensures better intestinal absorption by passive diffusion (acetoxy ester groups of cefuroxime-axetil, cefpodoxime-proxetil). Esterases in the intestinal mucosa release the active molecule.

For precise antibiotic therapy with cephalosporins, it is important to consult the ANSM NIH website or specialist dictionaries, in order to identify the sensitive germs.

Other molecules

Clavulanic acid is produced by *Streptomyces clavuligerus* and has an oxapenam structure with two asymmetric carbons [2R, 5R]. Its inhibitory activity is mainly against beta-lactamases from GRAM + and GRAM – germs. It has reduced activity against cephalosporinases. This compound has low intrinsic therapeutic activity and is used in combination with other penicillins, as already indicated:

– clavulanic acid + Amoxicillin;

– clavulanic acid + Ticarcillin.

Thienamycin, a member of the thia-carbapenem family, isolated from *Streptomyces cattleya*, has broad-spectrum antibacterial activity. It inhibits beta-lactamases. It is active against microorganisms that are resistant to penicillins and cephalosporins. Thienamycin is unstable and sensitive to oxidation, hence the synthesis of *Imipenem* (N-formimidoyl thienamycin derivative), which is stable and justifies its form of use. However, Imipenem is intensively metabolized in the body by dehydropeptidase I (DHP-1; EC 3), a renal enzyme, which inactivates it. Under these conditions, this derivative is also used in combination with *Cilastatin*, a structural analogue of Imipenem. This inhibits DHP-1 and improves activity compared with the natural product. Imipenem has antibacterial activity against GRAM + and GRAM – aerobic germs and acts primarily on Enterobacteriaceae. The main targets are *Pseudomonas aeruginosa*, Acinetobacter, Haemophilus, Neisseria and Staphylococci.

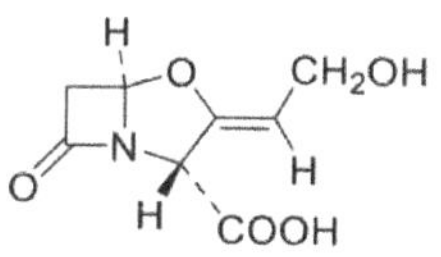

Clavulanic acid

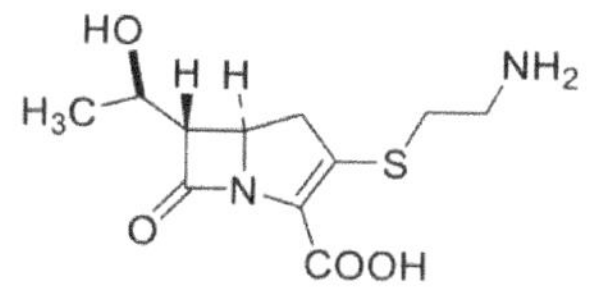

Thienamycin

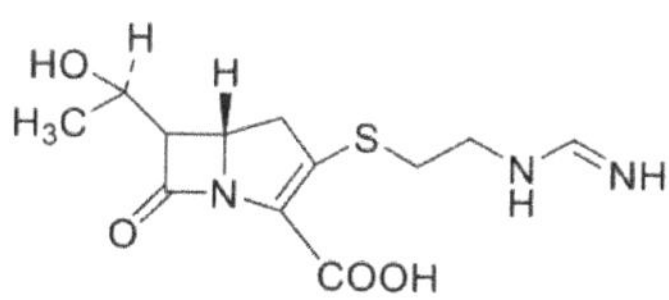

Imipenem

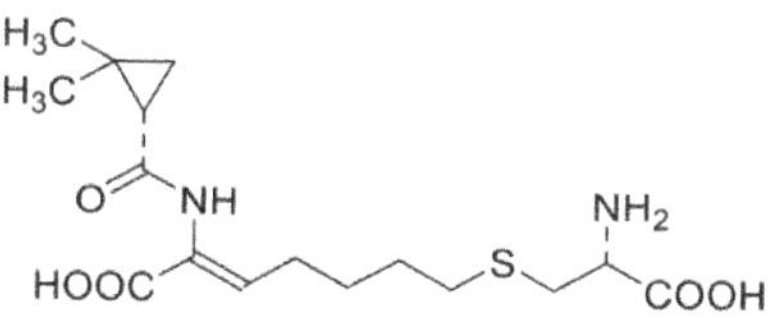

Cilastatin

Ertapenem is a Carbapenem derivative with a 2-aminocarbonylpyrrolidinyl-4-thio chain. This broad-spectrum antibiotic is stable against renal dehydropeptidase-1 and most beta-lactamases (Brands et al. 2002). It is used parenterally to treat infections caused by susceptible germs.

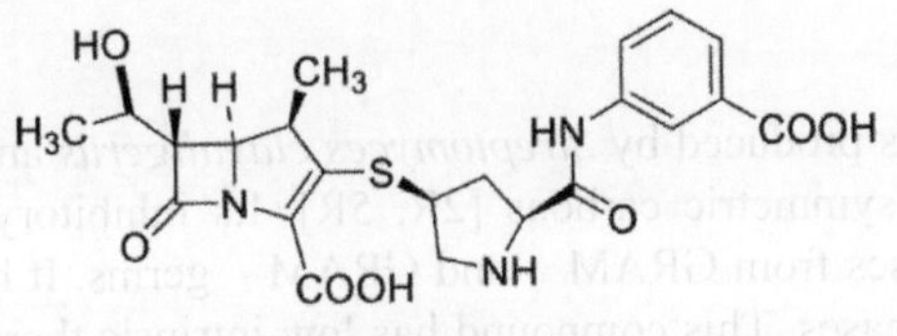

Ertapenem

Sulbactam and its derivative, *Sultamicillin*, as well as *Tazobactam* all contain a sulfone group.

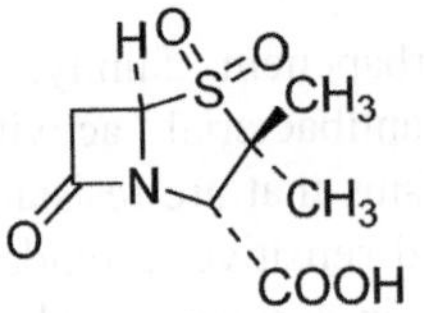

Sulbactam

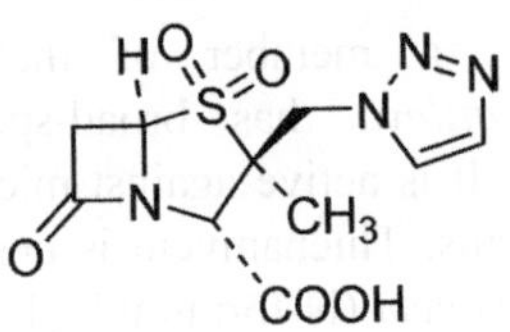

Tazobactam

Sulbactam is an irreversible inhibitor of extended-spectrum beta-lactamases (ESBL) in GRAM + and GRAM – bacteria. With low antibacterial activity, it is used in combination with penicillins and cephalosporins, whose antibacterial properties are improved. This helps to combat beta-lactamases.

Sultamicillin corresponds to the structure of a precursor of Sulbactam (a beta-lactamase inhibitor) and Ampicillin, which maintains antibacterial activity against GRAM + and GRAM – cocci and bacilli.

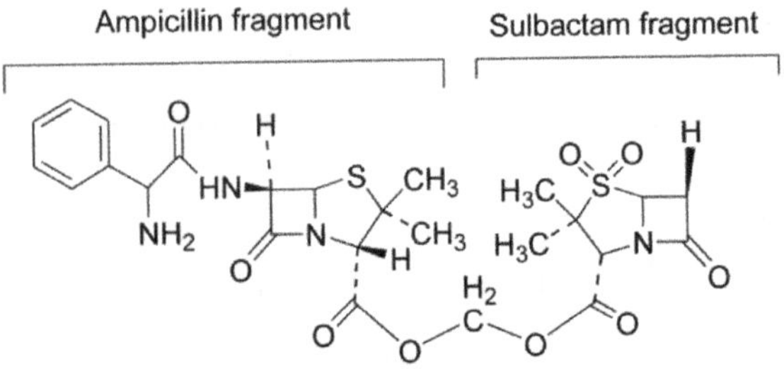

Sultamicillin is more bioavailable than any of the other components (80%). The compound is used to treat respiratory, ENT, genital, gynecological, skin and joint infections caused by sensitive germs.

Tazobactam is a Sulbactam derivative with a triazolyl substituent in one of its methyl groups. This compound is an irreversible inhibitor of extended-spectrum beta-lactamases (ESBL). It is used in combination: Tazobactam + Piperacillin.

Aztreonam is a monobactam with a beta-lactam ring structure, but differs from previous inhibitors by the absence of a link with another cyclic derivative.

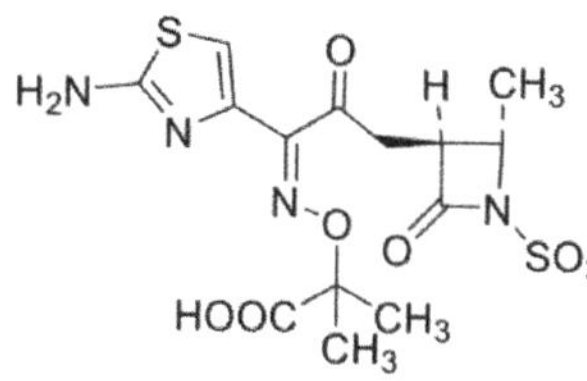

Aztreonam

It has good stability against beta-lactamases, but its action induces deformation of the bacterial wall by binding to proteins, resulting in filamentous forms. Its spectrum of activity is limited to bacilli and aerobic GRAM – cocci: Enterobacteriaceae (*E. coli*, Salmonella, Shigella, Enterobacter, Klebsiella, Serratia, Citrobacter, Proteus, *Haemophilus influenzae*, *Pseudomonas aeruginosa*, Neisseria and Moraxella).

Avibactam is a compound whose structure differs from previous beta-lactamase inhibitors in the absence of the beta-lactam residue. It has irreversible broad-spectrum beta-lactamase inhibitory activity and can be used exclusively by intravenous injection. Precursors have been produced with a view to obtaining orally active derivatives. The precursor, whose structure includes a neopentyl substituent, makes oral administration possible (Gordon et al. 2018).

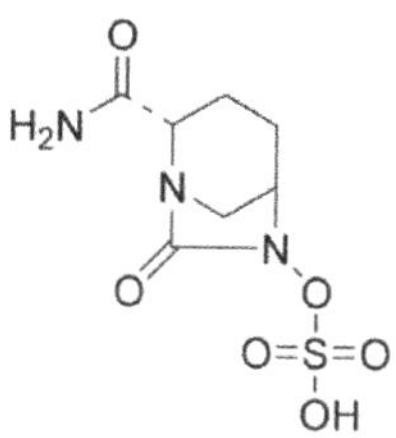

Avibactam

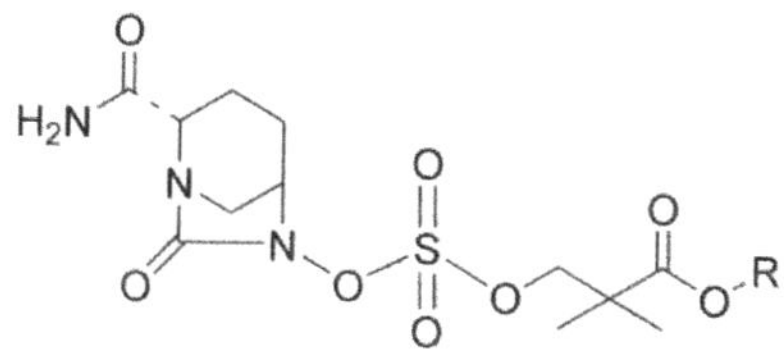

Avibactam precursor (prodrug)

The Avibactam + Ceftazidime combination (contains Ceftazidime (a third-generation cephalosporin) and Avibactam (a beta-lactamase inhibitor)), providing activity against resistant germs. It is specified for the treatment of intra-abdominal or urinary tract infections and nosocomial pneumonia.

ASSESSMENT.– *The search for inhibitors of bacterial beta-lactamases and enzymes involved in the synthesis of the bacterial wall is still very extensive, as indicated in section 2.2.2.4.2. They are fully fledged inhibitors of bacterial enzymes and/or act indirectly to boost the action of penicillins and cephalosporins degraded by beta-lactamases and cephalosporinases. There are numerous combinations of the two types of inhibitor. Some have been discussed.* Recently, Enmetazobactam, a new beta-lactamase inhibitor, was also introduced.

2.2.2.5. *UDP-N-acetylglucosamine 1-carboxyvinyltransferase (EC 2)*

This enzyme encoded by the murA gene, also known as enoyl-pyruvyl-transferase, catalyzes the first stage in peptidoglycan biosynthesis. It transfers the enolpyruvate part of phosphoenolpyruvate to UDP-N-acetylglucosamine.

Fosfomycin inhibits this enzyme by behaving like an analogue of the substrate phosphoenolpyruvate.

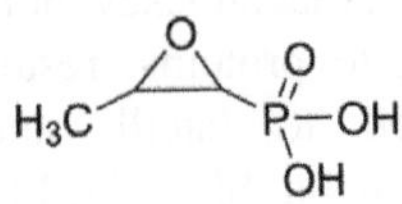

Fosfomycin

In order to act, the antibiotic must penetrate the bacterial cytoplasm. This requires active transport. This molecule does not have good parameters for intestinal absorption. It is often administered parenterally. This antibiotic is active on aerobic GRAM + bacteria (*Enterococcus faecalis*, *Streptococcus pneumoniae* and Staphylococci meti-R or -S), aerobic GRAM – bacteria (Citrobacter, Enterobacter, *Escherichia coli*, *Haemophilus influenzae*, Klebsiella, Pasteurella, *Proteus vulgaris* and *mirabilis* or *Pseudomonas aeruginosa*, etc.). Fosfomycin also acts on Fusobacterium.

2.2.2.6. *Bacterial urease (EC 3)*

This is a hydrolase capable of transforming urea into carbon dioxide and ammonia following the addition of water. This enzyme has been characterized as a virulence factor in many bacterial species, including *Proteus mirabilis*, *Staphylococcus saprophyticus* and *Helicobacter pylori*. In *Helicobacter pylori*, for example, urease plays a central role in its metabolism and virulence. It is essential for colonization of the gastric mucosa by microorganisms.

Acetohydroxamic acid is an inhibitor of this enzyme. It acts as an analogue of its substrate, urea. Its action is often considered insufficient, but it is used in some countries. The antibiotic is responsible for significant adverse effects, such as cancer induced by a metabolite, acetamide.

2.2.2.7. *Dihydrofolate synthetase (EC 6) and dihydrofolate reductase (EC 1)*

Dihydrofolate synthetase is an essential enzyme that contributes to the production of dihydrofolate. Folate is vitamin B9, an essential micronutrient.

Dihydrofolate reductase produces tetrahydrofolate from dihydrofolate and NADPH. Tetrahydrofolate is involved in the transport of monocarbon radicals.

Antibacterial sulfonamides, which have been widely used since they were first introduced as therapeutic agents in the 1940s, have seen their numbers decline considerably with the development of resistance and the introduction of antibiotics. Their number is currently limited to a few representatives, that are mainly used in combination against sensitive germs. They were discovered during the search for antistreptococcal and antistaphylococcal substances. After assessing the activity of over a thousand azo dyes, G. Domagk (1935) discovered the antistreptococcal activity of prontosil in vivo in mice. This substance had no antibacterial activity in vitro.

Subsequent work at the Institut Pasteur (Tréfouël et al. 1935) showed that of the two fragments formed by an azoreductase from this azo dye, only the sulfonamide fragment was responsible for antibacterial activity. Bacteria biosynthesize the folic acid that they require in order to grow. They are unable to take it from the host due to the impermeability of their membrane. Sulfonamides are structural analogues of the p-aminobenzoic acid (APAB) required for the biosynthesis of folic acid, a growth factor for bacteria.

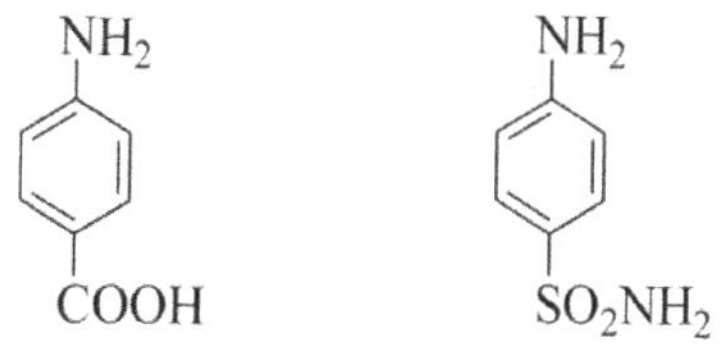

p-aminobenzoic acid **p-aminobenzenesulfamide**

The second stage in this synthesis pathway, catalyzed by dihydrofolate synthetase, involves the binding of L-glutamate to dihydropteroate. Instead of dihydropteroate, an antimetabolite with a sulfonamide fragment inhibits the enzyme. By taking the place of the APAB fragment, sulfonamides block the formation of dihydrofolic acid. This inhibition prevents the synthesis of dihydrofolate, which is necessary for the synthesis of nucleic acids that are essential for bacterial multiplication (Figure 2.20). Sulfonamides thus have bacteriostatic activity. Bacteria that no longer have dihydrofolic acid are unable to subsequently produce the tetrahydrofolic acid they need for growth. Sulfonamides can cause kidney and hematopoietic disorders. They have a very broad spectrum of action, including GRAM + and GRAM – bacteria, anaerobes, Chlamydophila, toxoplasmas and cocci. Pseudomonas, mycoplasma and spirochetes are resistant. Their clinical use may be limited due to acquired resistance.

It should be noted that sulfonamides have another target, dihydropteroate synthetase, the first enzyme in the metabolic pathway. This enzyme will be discussed in section 2.2.2.8.

In order to achieve more powerful bactericidal activity than the bacteriostatic action of sulfonamides, treatments combine sulfonamides with an antifolinic (*Trimethoprim, Pyrimethamine*).

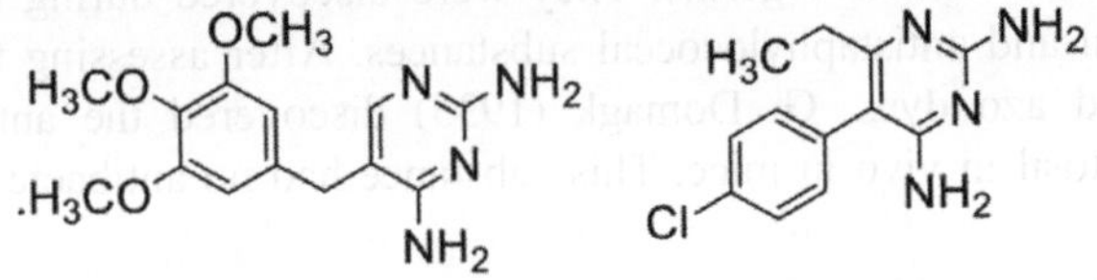

Trimethoprim **Pyrimethamine**

Antifolates inhibit dihydrofolate reductase and thus prevent the formation of 5,6,7,8-tetrahydrofolic acid (tetrahydrofolate), which is necessary for the formation of purines, and consequently for the biosynthesis of DNA, which is essential for bacterial multiplication (Figure 2.20). Antifolates are active against GRAM + and GRAM –, but not anaerobes. Their action against mycoplasmas and certain Chlamydophila and Pseudomonas is negligible.

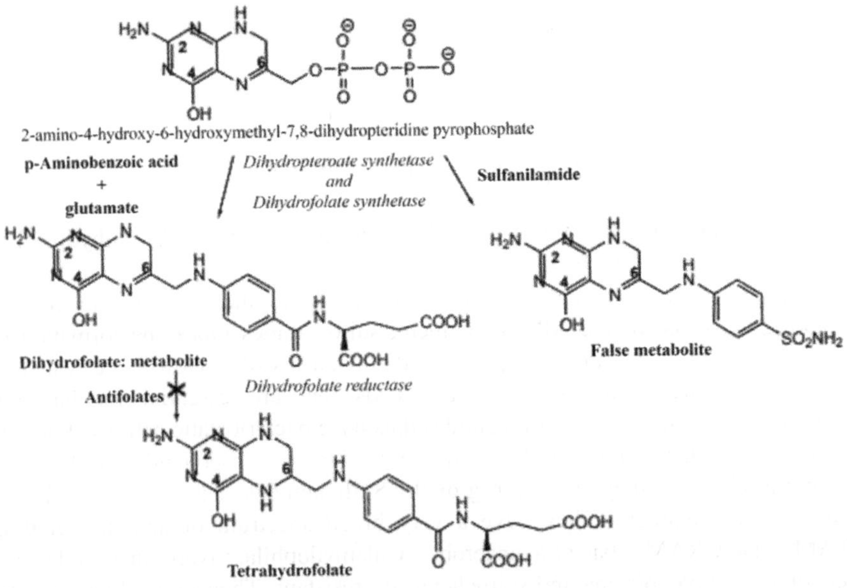

Figure 2.20. *Antibacterial action of a sulfonamide*

Under these combination conditions, the antibacterial spectrum is extended to GRAM + and GRAM – resistant germs (Cockerill and Edson 1991) (Figure 2.21). However, this combination also leads to the formation of bacterial resistance (Huovinen 2001).

Combining a sulfonamide that inhibits dihydrofolate synthetase (*Sulfame-thoxazole*), which prevents the formation of dihydrofolate, with an inhibitor of dihydrofolate reductase (Trimethoprim) improves antibacterial activity and makes it bactericidal. The use of sulfonamide alone would only lead to a less effective bacteriostatic effect.

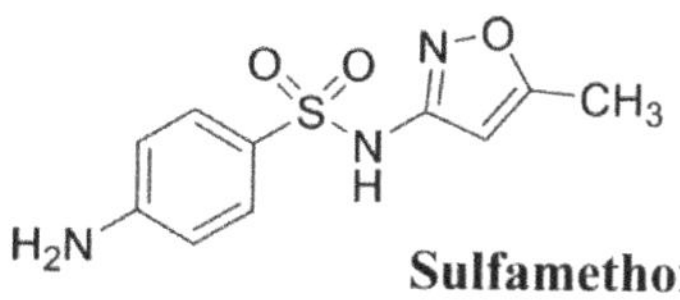

Sulfamethoxazole

Other combinations of sulfonamides and antifolinics are known, including *Sulfadoxine* + Pyrimethamine.

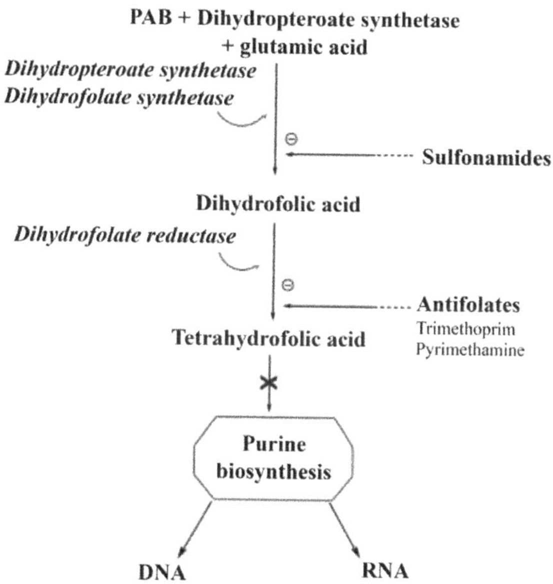

Figure 2.21. *Combination of a sulfonamide and an antifolate agent*

NOTE.– Methotrexate's inhibition of folates in mammals involved in the biosynthesis of purines and thymidine is also used in the treatment of certain cancers (see section 3.5.2).

2.2.2.8. *Dihydropteroate synthase (EC 2)*

This enzyme is involved in the folate synthesis pathway. It binds 6-hydroxy-methyldihydropterin pyrophosphate and para-amino benzoic acid to form 7,8-dihydropteroate and pyrophosphate. Molecules in the sulfone family (such as *Dapsone*) or the sulfonamide family inhibit this enzyme. Dapsone acts in particular against Hansen's bacillus (*Mycobacterium leprae*). This antibiotic is used to treat leprosy and certain dermatoses.

2.3. Fungal targets

2.3.1. *1,3-beta-glucan synthase (EC 2)*

This is a multimeric membrane enzyme that is specific to fungi. It catalyzes the polymerization of glucose molecules linked by β–1,3 bonds. The last glucose, in activated form (UDP-glucose), is added to the polymer being formed. Glucan forms part of the wall of fungi, a binding and protective structure that is very important for many filamentous fungi and yeasts. Inhibition of this enzyme weakens the fungus wall, increases its permeability and leads to loss of cell content, resulting in lysis and death.

Caspofungin is a semi-synthetic lipopeptide (echinocandin), produced from a fermentation product of *Glarea lozoyensis* (a species of fungus). This molecule inhibits the synthesis of β–(1,3)-D-glucan. This glucan is absent from mammalian cell composition. The fungicidal activity of this compound has been demonstrated on candida (the fungi responsible for candidiasis, which accounts for almost 80% of invasive fungal infections).

Other studies show that exposure of aspergillus (15% of invasive fungal infections, known as aspergillosis) to caspofungin leads to lysis and death of the ends of mycelial filaments, as well as the sites where cell growth and division take place. It is a non-competitive inhibitor of the enzyme.

Anidulafungin is another drug that inhibits this enzyme, prepared by hemisynthesis from a fermentation product of *Aspergillus nidulans*.

Micafungin is prepared by hemisynthesis from a fermentation product of *Coleophoma empetri*.

Echinocandins, such as Anidulafungin, Caspofungin and Micafungin, are cyclic lipopeptides with antifungal activity, which act by blocking the biosynthesis of 1,3-beta-D-glucan.

NOTE.– Indications and use: serious invasive fungal infections caused by candida and aspergillus in patients who cannot tolerate other antifungal agents or who are resistant to other treatments (Quindós et al. 2019).

2.3.2. *Squalene mono-oxygenase (EC 1)*

Squalene mono-oxygenase is an enzyme that catalyzes the conversion of squalene to 2,3-epoxysqualene in the ergosterol biosynthesis pathway. It is an interesting target for drugs used to inhibit the growth of pathogenic fungi. This same enzyme can be targeted in human pathologies linked to hypercholesterolaemia.

Tolnaftate is a synthetic narrow-spectrum antifungal inhibitor of this enzyme. In vivo, it has antifungal activity against dermatophytes, *Aspergillus niger* and *Pityrosporum orbiculare.*

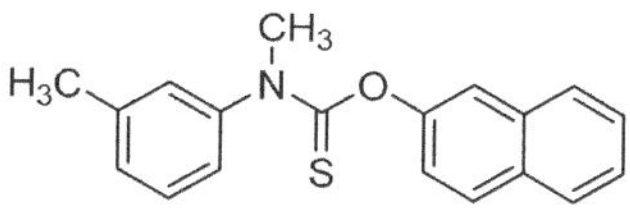

Tolnaftate

2.3.3. *14-sterol demethylase (EC 1)*

This enzyme, also known as cytochrome P450 51, catalyzes 14-alpha demethylation, a biochemical reaction which forms part of the sterol synthesis pathway in eukaryotes. *Bifonazole* is an imidazole antifungal that is active on a broad spectrum, including dermatophytes, yeasts (*Malassezia furfur*), molds and GRAM + bacteria such as Corynebacterium and Staphylococcus.

Bifonazole

2.3.4. *Thymidylate synthase (EC 2)*

The enzyme is involved in the synthesis of thymidine and the formation of DNA molecules. It catalyzes the transfer of a monocarbon radical onto deoxy-UMP to obtain deoxy-TMP.

Pentamidine acts on the *Pneumocystis carinii* enzyme.

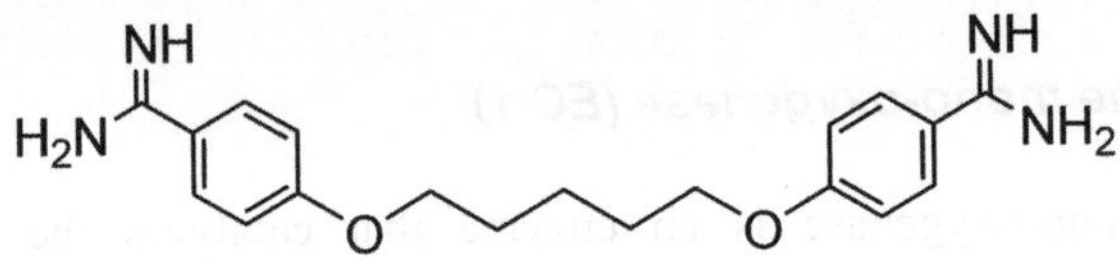

Pentamidine

2.4. Parasite targets

2.4.1. *Ornithine decarboxylase (EC 4)*

This is an enzyme that enables ornithine to be converted into putresin, which is the precursor of polyamines. *Trypanosoma brucei* is a species of trypanosome that possesses this enzyme. This parasite is responsible for sleeping sickness, also known as African trypanosomiasis.

Eflornithine irreversibly inhibits ornithine decarboxylases.

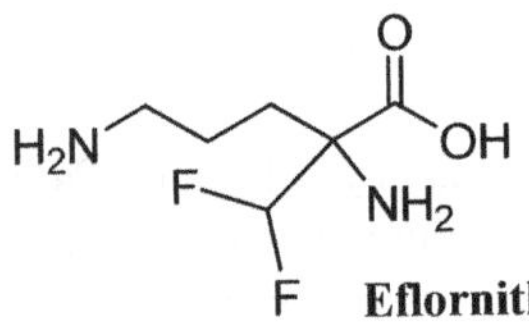

Eflornithine

NOTE.– Other drugs are used against these pathogens, such as Pentamidine for *Trypanosoma brucei gambiense* and *Suramine* for *Trypanosoma brucei rhodesiense.*

These two compounds appear to have a trypanocidal effect, either by inhibiting DNA synthesis by blocking thymidylate synthase or by binding to transfer RNA.

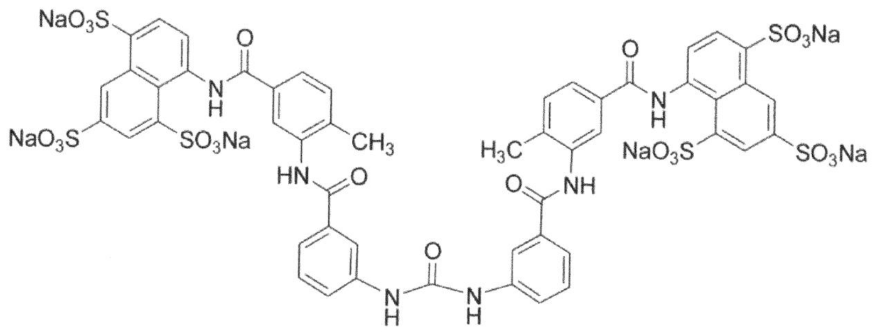

Suramin sodium

2.4.2. *Heme polymerase (EC 2)*

Plasmodium falciparum is a parasite that is responsible for human malaria. It causes almost 500,000 deaths a year worldwide. The enzyme polymerizes heme nuclei with the aim of multiplying the pathogenic microorganism.

The parasite's proteases hydrolyze the protein part of hemoglobin to obtain the amino acids essential for the parasite's survival. Simultaneously, the individual hemes released are toxic to the parasite. The infectious agent's heme polymerase enables the ferriprotoporphyrin IX (heme) units to be combined, making the polymer formed non-toxic to the plasmodium.

Chloroquine binds to ferriprotoporphyrin IX, producing a substance that is toxic to the parasite. This compound is thought to play a significant role in preventing the development of plasmodium in red blood cells. Chloroquine is not limited to this action, and other additional mechanisms may be involved.

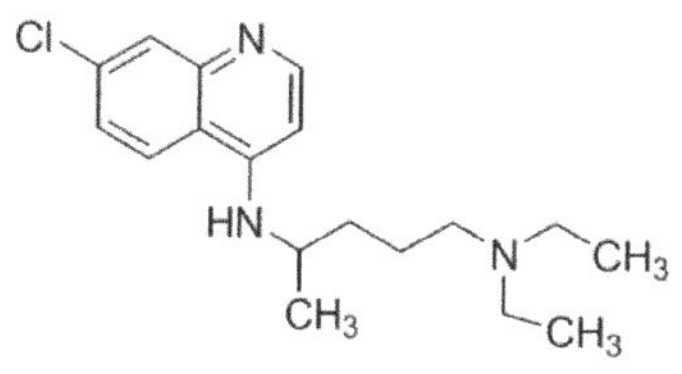

Chloroquine

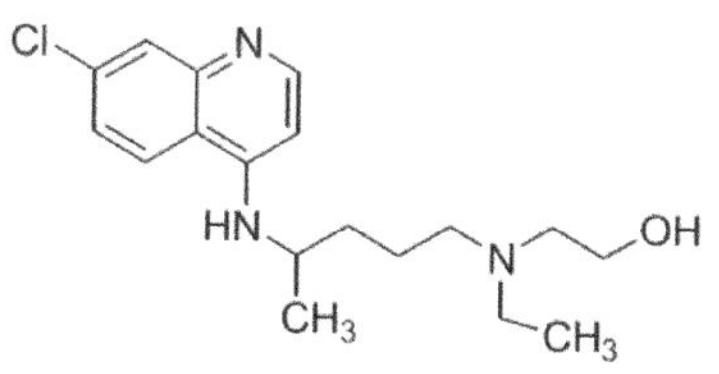

Hydroxychloroquine

2.5. References

Aristoff, P.A. (1998). Dihydropyrone sulfonamides as a promising new class of HIV protease inhibitors. *Drugs Future*, 23, 995–999.

Barreteau, H., Koval, A., Boniface, A., Sova, M., Gobec, S., Blanot, D. (2008). Cytoplasmic steps of peptidoglycan biosynthesis. *FEMS Microbiol Rev*, 32, 168–207.

Bourinbaiar, A.S., Tan, X., Nagorny, R. (1993). Effect of the oral anticoagulant, warfarin, on HIV-1 replication and spread. *AIDS*, 7, 129–213.

Brands, K.M.J., Jobson, R.B., Conrad, K.M., Williams J.M., Pipik B., Cameron M., Davies A.J., Houghtonn P.G., Ashwood M.S. et al. (2002). Efficient one-pot synthesis of the 2-aminocarbonylpyrrolidin-4-thio-containing side chain of the new broad-spectrum carbapenem antibiotic ertapenem. *Org Chem*, 67(14), 4771–4777.

Cockerill, F.R. and Edson, R.S. (1991). Trimethoprim-sulfamethoxazole. *Mayo Clin Proc*, 66(12), 1260–1269.

Drawz, S.M. and Bonomo, R.A. (2010). Three decades of beta-lactamase inhibitors. *Clin Microbiol Rev*, 23, 160–201.

Duguide, J.P. (1946). The sensitivity of bacteria to the action of penicillin. *Edinb Med J.*, 53, 401.

Floss, H.G. and Yu, T.W. (2005). Rifamycin-mode of action, resistance, and biosynthesis. *Chem Rev*, 105, 621–632.

Huovinen, P. (2001). Resistance to trimethoprim-sulfamethoxazole. *Clin Infect Dis*, 32(11), 1608–1614.

Imming, P., Sinning, C., Meyer, A. (2006). Drugs, their targets and the nature and number of drug targets. *Nat Rev Drug Discov*, 5(10), 821–834.

Kirkiacharian, S. (2007). Chimie médicinale. Structure et activité du médicament. *Techniques de l'Ingénieur*.

Kirkiacharian, S. (2010). *Guide de chimie médicinale et médicaments*. Lavoisier, Paris.

Kirkiacharian, S., Thanh Thuy, D., Sicsic, S., Bakhchinian, R., Kurkjian R., Tonnaire T. (2002). Structure-activity relationships of some 3- substituted-4-hydroxycoumarins as HIV-1 protease inhibitors. *Il Farmaco*, 57, 703–708.

Lee, J.S., Calmy, A., Andrieux-Meyer, I., Ford, N. (2012). Review of the safety, efficacy, and pharmacokinetics of elvitegravir with an emphasis on resource-limited settings. *HIV AIDS (Auckl)*, 4, 5–15.

Lu, X., Lv, M., Huang, K., Ding, K., You, K. (2012). Pharmacophore and molecular docking guided 3D-QSAR study of bacterial enoyl-ACP reductase (FabI) Inhibitors. *Int J Mol Sci*, 13(6), 6620–6638.

Gordon, E.M., Duncton, M.A.J., Gallop, M.A. (2018). Orally absorbed derivatives of the beta-lactamase inhibitor avibactam. design of novel prodrugs of sulfate containing drugs. *J Med Chem*, 61, 10340–10344.

Park, J.T. and Strominger, J.L. (1957). Mode of action of penicillin. *Science*, 125, 99–101.

Quindós, G., Gil-Alonso, S., Marcos-Arias, C. (2019). Therapeutic tools for oral candidiasis: Current and new antifungal drugs. *Med Oral Patol Oral Cir Bucal*, 24(2), e172–e180.

Ramanathan, S., Mathias, A.A., German, P., Kearney B.P. (2011). Clinical pharmacokinetic and pharmacodynamic profile of the HIV integrase inhibitor elvitegravir. *Clin Pharmacokinet*, 50(4), 229–244.

Rask-Andersen, M., Almén, M.S., Schiöth, H.B. (2011). Trends in the exploitation of novel drug targets. *Nat Rev Drug Discov*, 10(8), 579–590.

Robertson, J.G. (2005). Mechanistic basis of enzyme-targeted drugs. *Biochemistry*, 44(15), 5561–5571.

Sauvage, E., Kerff, F., Terrak, M., Ayala, J.A., Charlier, P. (2008). The penicillin-binding proteins: Structure and role in peptidoglycan biosynthesis. *FEMS Microbiol Rev*, 32(2), 234–258.

Spratt, B.G. (1983). Penicillin-binding proteins and the future of beta-lactam antibiotics. The Seventh Fleming Lecture. *J Gen Microbiol*, 129(5), 1247–1260.

Thaisrivongs, S., Tomich, P.K., Watenpaugh, K.D., Chong, K.-T., Howe, W.J., Yang, C.-P., Strohbach, J.W., Turner, S.R., McGrath, J.P., Bohanon, M. (1994). Structure-based design of HIV protease inhibitors: 4-hydroxycoumarins and 4-hydroxy-2-pyrones as non-peptidic inhibitors. *J. Med. Chem.*, 37, 3200–3204.

Thaisrivongs, S., Skulnick, H.I., Turner, S.R., Strohbach, J.W., Tommasi, R.A., Johnson, P.D., Aristof, P.A., Judge, T.M., Gammill, R.B., Morris, J.K. et al. (1996). Structure-based design of HIV protease inhibitors: Sulfonamide- containing 5,6-dihydro-4-hydroxy-2-pyrones as non-peptidic inhibitors. *J Med Chem*, 39, 4349–4355.

Tipper, D.J. (1985). Mode of action of beta-lactam antibiotics. *Pharmacol Ther*, 27(1), 1–35.

Tréfouël, J., Tréfouël, T., Nitti, F., Bovet, D. (1935). Activité du p.aminophénylsulfamide sur l'infection streptococcique expérimentale de la souris et du lapin. *Comptes Rendus Soc Biol.*, 120, 756–758.

Triola, G., Wetzel, S., Ellinger, B., Koch, M.A., Hübel, K., Rauh, D., Waldmann, H. (2009). ATP competitive inhibitors of D- alanine–D-alanine ligase based on protein kinase inhibitor scaffolds. *Bioorg & Med Chem*, 17(3), 1079–1087.

3

Targeted Human Enzymes

3.1. Treatment via effectors

Given the large number of enzymes to be detailed in this chapter, a simplified classification, which is not intended to be set in stone, will be applied. This chapter takes into account the main field of application where the enzyme is targeted, as indicated below: ophthalmology, neurology, metabolism, endocrinology, cardiovascular system, immunology and oncology. In the future, the scope of research will enable us to extend the list of targeted enzymes and refine this classification.

3.1.1. *Ophthalmology*

3.1.1.1. *Carbonic anhydrase (EC 4)*

There are various carbonic anhydrase isoenzymes. *Dorzolamide* and *brinzolamide* are powerful carbonic anhydrase inhibitors. They are used in ophthalmology.

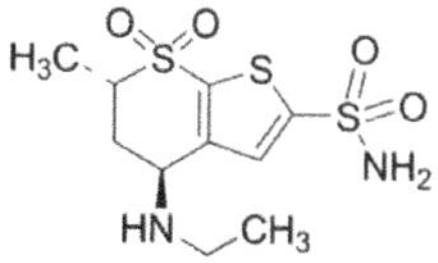

Dorzolamide

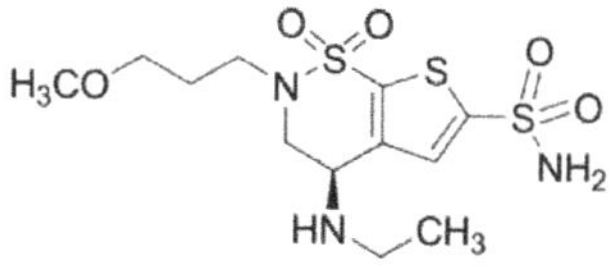

Brinzolamide

Acetazolamide has a wider range of uses, probably due to its action on several isoenzymes. It is recommended in ophthalmology for the treatment of metabolic alkalosis and following post-operative or traumatic edema.

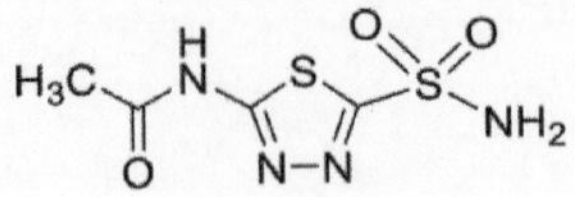

Acetazolamide

A great deal of research is carried out on carbonic anhydrase isoenzymes.

3.1.1.2. *Aldose reductase (EC 1)*

D-glucose metabolism in normoglycemic subjects involves the conversion of glucose into glucose-6-phosphate (G6-P) by hexokinase. G6-P enters glycolysis and contributes to the body's energy requirements. In diabetics, when glucose levels are high and hyperglycemia is prolonged, hexokinase becomes saturated, with sorbitol accumulating under the action of aldose reductase. The action of sorbitol dehydrogenase is insufficient to metabolize all the sorbitol formed into D-fructose (Figure 3.1).

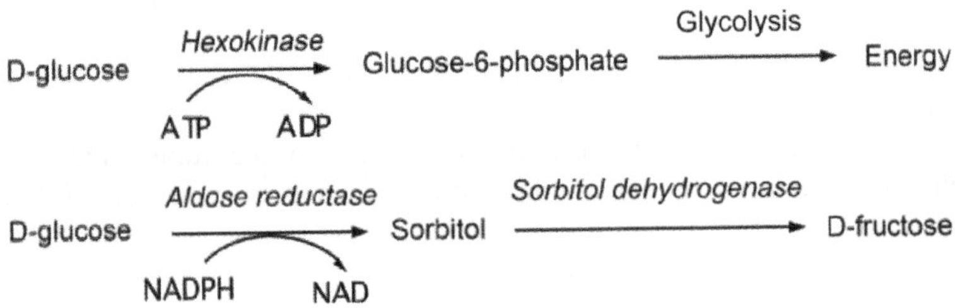

Figure 3.1. *Aldose reductase action*

Increased concentrations of sorbitol and fructose in the eye, nerve tissue and various organs are the cause of eye-related complications (retinopathies, cataracts, optic nerve damage and blindness), neurological complications (neuropathies), nephropathies and vasculopathies, hence the importance of aldose reductase inhibitors.

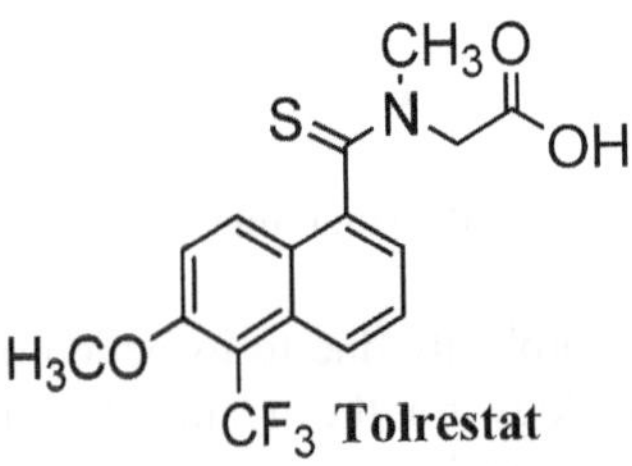

Tolrestat

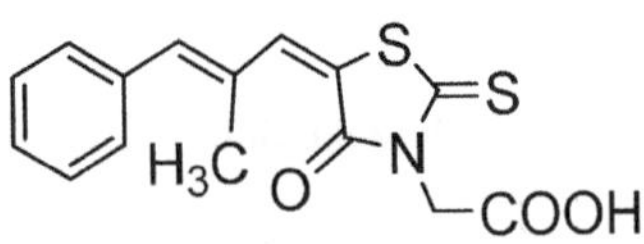

Epalrestat

Various inhibitors of this enzyme are known and marketed in some countries, including *Tolrestat* and *Epalrestat*.

Other substances, such as *Gossypol* and *Quercetin*, act on this catalyst.

3.1.2. *Neurology*

3.1.2.1. *Acetylcholinesterase (EC 3)*

The catalyst hydrolyzes acetylcholine into acetic acid and choline in postsynaptic cells that have bound the neurotransmitter to their specific receptor. This hydrolysis stops the signal generated by acetylcholine binding. An inhibitor of this enzyme prolongs the action of acetylcholine.

Acetylcholinesterase is targeted in Alzheimer's disease in patients with mild to moderate forms of the disease. Acetylcholine is in short supply in the synaptic clefts of the central nervous system. As this neurotransmitter plays a role in memory processes, its half-life needs to be increased. It should be noted that *Tacrine*, *Donepezyl*, *Galantamine* and *Rivastigmine* are acetylcholinesterase inhibitors that have been proposed in the treatment of mild to moderately severe forms of this disease.

Neostigmine is also an acetylcholinesterase inhibitor.

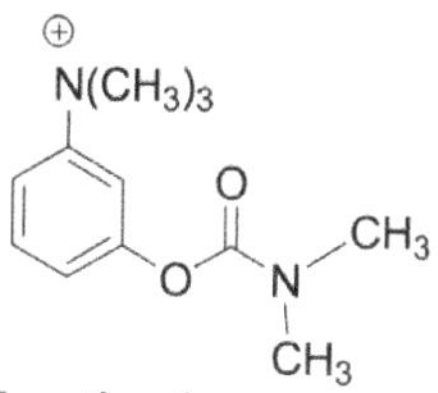

Neostigmine

This molecule is used for post-operative decurarization and myasthenia. This compound prolongs and enhances the muscarinic and nicotinic effects of acetylcholine. For the first condition, the molecule gradually lifts the effect of non-depolarizing curares on "blocked" neuromuscular junctions. For the second condition, it increases the intensity and rhythm of smooth muscle fiber contractions, promoting peristalsis, and normalizes the strength of striated muscle contractions (anti-myasthenic effect) by promoting nerve influx.

In sections 3.1.2.2–3.1.2.4, monoamine oxidase B (MAO-B), L-DOPA decarboxylase and catechol-O-methyltransferase (COMT) are grouped together for the treatment of Parkinson's disease.

This pathology is characterized by the destruction of dopaminergic neurons in the nigrostriatal region. This leads to a dopamine deficit in the striatum of the central nervous system, associated with hyperactivity of the cholinergic system, secondary to the loss of dopaminergic neurons. The symptoms are tremors, muscular rigidity (hypertonia) and loss of automated movement (akinesia).

The treatment of this disease aims to:

– inhibit peripheral degradation of Levodopa, to preserve its maximum availability for passage to the central nervous system;

– inhibit monoamine oxidase (MAO) and catechol-O-methyl-transferase (COMT), which are dopamine-catabolizing enzymes, to prevent degradation and loss of activity (Figure 3.2);

– administer dopamine agonists.

Dopamine administration is ineffective because its polarity does not allow it to cross the blood–brain barrier, unlike its precursor, Levodopa, which is converted into dopamine by DOPA decarboxylase (Figure 3.2).

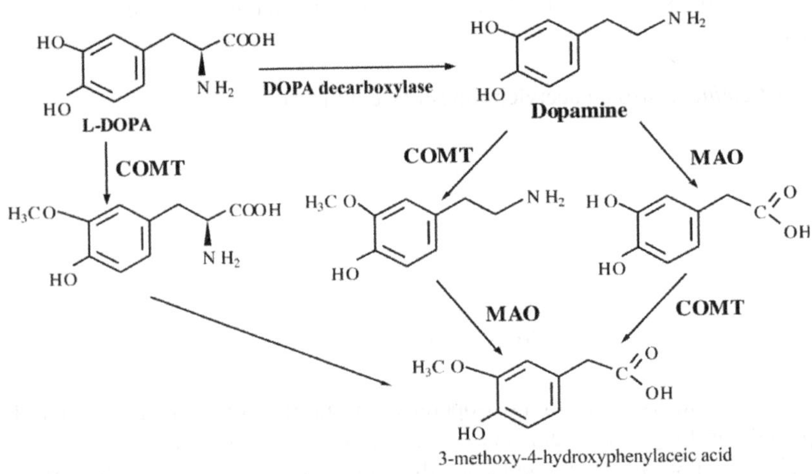

Figure 3.2. *Biosynthesis and catabolism of Levodopa and dopamine*

3.1.2.2. *Monoamine oxidases A and B (EC 1)*

The discovery of antidepressant properties during treatment with the anti-tuberculosis drug Isoniazid led to the discovery of the therapeutic class of monoamine oxidase inhibitors (MAOIs). Two enzyme type inhibitors were subsequently identified: type A (MAOI-A) and type B (MAOI-B). Both enzymes are monoamine catabolizing oxidoreductases. The action of monoamine oxidases (MAOs) results in the oxidative deamination of physiological monoamines according to the reaction shown in Figure 3.3.

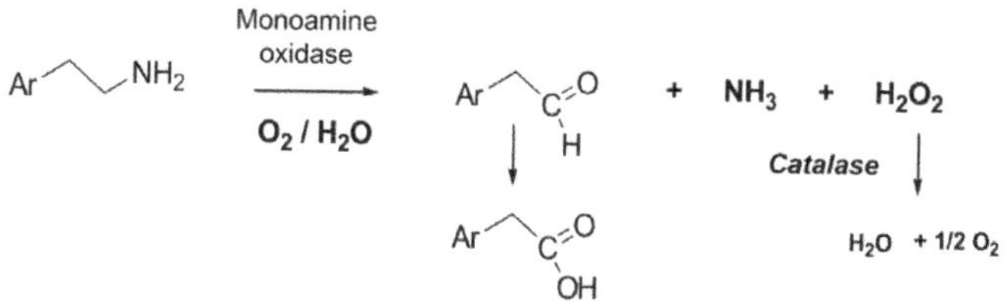

Figure 3.3. *Action of monoamine oxidases (MAO)*

MAO inhibition increases the concentration of neurotransmitters in the synaptic cleft. Subsequent research led to derivatives inspired by the hydrazide structure of Isoniazid, such as *Iproniazid* and *Nialamide*, used as anti-tuberculosis drugs.

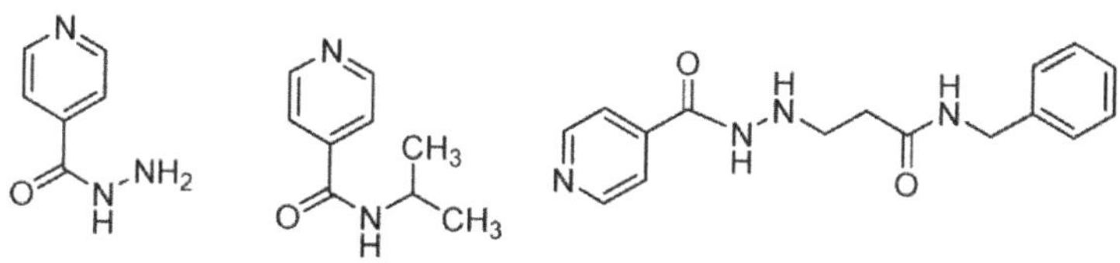

Isoniazid **Iproniazid** **Nialamide**

These are first-generation MAOIs with no selectivity for either monoamine oxidase A or B, and with dopamine and tyramine as common substrates.

Iproniazid (phosphate) and Nialamide are non-selective MAOIs. They act as non-selective MAO A and B inhibitors. As a result, they block amine catabolism, enhancing their effects at different levels:

– Cerebral: antidepressant effect.

– Cardiac: analgesic action (reduced oxygen consumption).

– Hepatic: inhibition of detoxification enzymes (hepatotoxicity), with the risk of potentiation of many drugs when co-administered (analgesics, psychotropics, antihypertensives, etc.). Side effects include constipation, dry mouth and difficulty in micturition (ganglioplegic and parasympatholytic effects).

– The action of non-selective MAOIs also extends to intestinal MAO-A, with normal degradation of dietary tyramine. This lack of MAO-A inhibition can lead to a severe hypertensive flare-up known as the "cheese effect" when fermented cheeses and foods rich in tyrosine and tyramine are consumed. These molecules are precursors in the biosynthesis of hypertensive catecholamines, blocking their degradation despite their high absorption.

Subsequent work led to the discovery of selective substrates for each of the two types of MAO, whose inhibition led to derivatives with different biological activities:

– MAO-A leads to preferential deamination of substrates such as norepinephrine, epinephrine and serotonin. It is irreversibly inhibited by *Clorgyline*;

– MAO-B has *Tryptamine*, *Benzylamine* and *Phenethylamine* as specific substrates. It is irreversibly inhibited by *Deprenyl* (see below).

MAOIs also induce adverse effects with many drugs, especially those with psychotropic activity and H1 antihistamines.

Second-generation selective MAOIs with different substrates offer reversible selective effects. They are more manageable therapeutically and have fewer side effects. *Toloxatone* and *Moclobemide* are used as antidepressants. Their action results from the inhibition of MAO-A and enzymatic catabolism of norepinephrine, serotonin and, to a lesser extent, dopamine in the central nervous system. They are used to treat depressive states, melancholic episodes and manic-depressive psychosis.

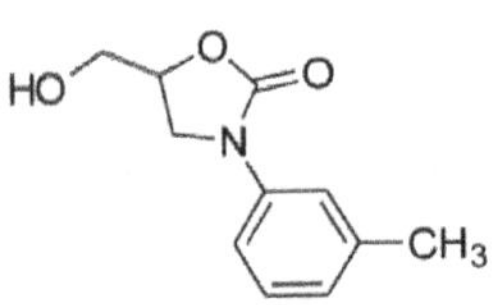

Toloxatone

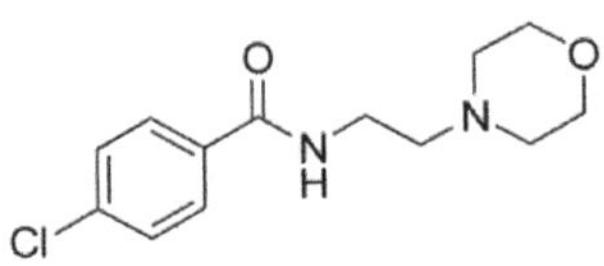

Moclobemide

Although MAO-A inhibitors do not interfere with the breakdown of exogenous tyramine, a tyramine-rich diet is not recommended during treatment.

MAO-B inhibitors are used in the treatment of Parkinson's disease. Two MAO-B inhibitors, shown below, are *Selegiline* and *Rasagiline*, which inhibit L-DOPA degradation.

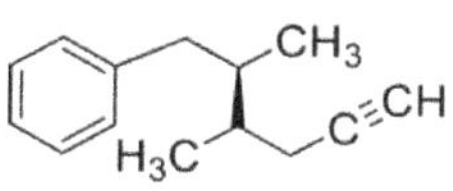

Selegiline

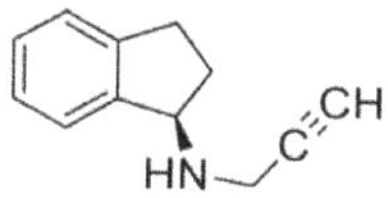

Rasagiline

3.1.2.3. *L-amino aromatic acid decarboxylase (EC 4)*

Aromatic L-amino acid decarboxylase, also known as L-DOPA decarboxylase, primarily catalyzes the decarboxylation of L-DOPA to form the catecholamine dopamine.

Inhibition of DOPA decarboxylase involves *Carbidopa* and *Benserazide,* administered simultaneously with *Levodopa.* This combination prevents the peripheral degradation of Levodopa, enabling better bioavailability in the brain.

The two combinations are Carbidopa + Levodopa and Benserazide + Levodopa. They are also characterized by a reduction in peripheral side effects. Treatment becomes more effective, while requiring lower doses.

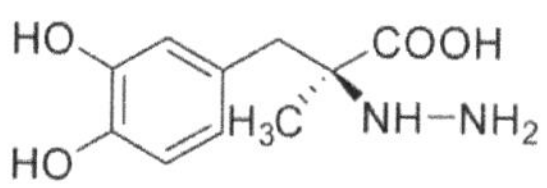

Carbidopa

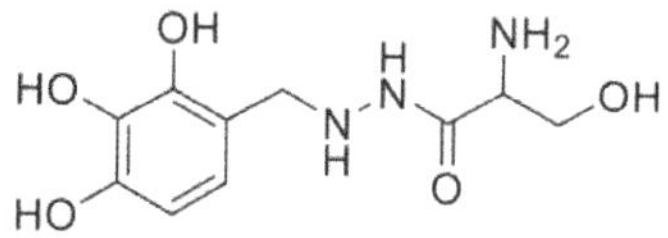

Benserazide

3.1.2.4. *Catechol-O-methyltransferase (EC 2)*

This enzyme, abbreviated as COMT, initiates the degradation of catecholamines by introducing a methyl group onto the structure, thanks to the cofactor S-adenosyl methionine (SAM). Inhibiting the enzyme results in longer half-lives for catecholamines and their derivatives. *Entacapone* is a specific, reversible inhibitor of this enzyme, as is *Tolcapone.*

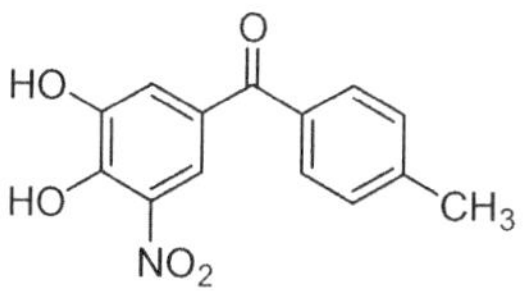

Tolcapone

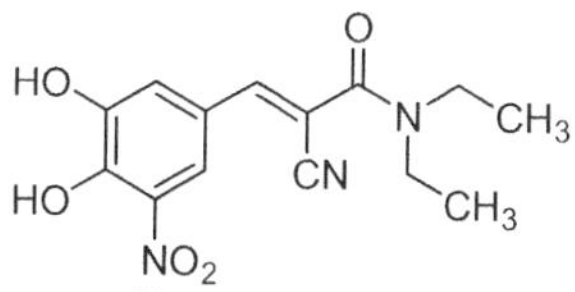

Entacapone

They mainly act at the peripheral level. They reduce the biotransformation of Levodopa into 3-O-methyldopa by inhibiting the enzyme. This increases the amount of Levodopa available in the brain. As with L-DOPA decarboxylase inhibitors,

COMT inhibitors are administered with Levodopa to increase brain bioavailability and avoid premature degradation.

3.1.2.5. *Succinate semi-aldehyde dehydrogenase (EC 1)*

Succinate semi-aldehyde dehydrogenase transforms succinic semi-aldehyde into succinate in the presence of NAD+ (Figure 3.4). *Valproic acid*-type drugs targeting this enzyme are used in the treatment of epilepsy.

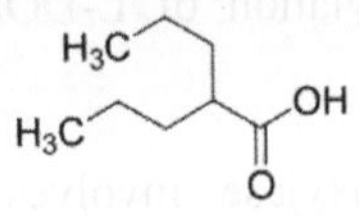

Valproic acid

3.1.2.6. *GABA transaminase (EC 2)*

This catalyst is responsible for the degradation of gamma-aminobutyric acid (GABA). 4-Aminobutyric acid is an inhibitory neurotransmitter that acts on three types of receptor. The GABA-A and GABA-C receptors are located in the brain, while the GABA-B receptor is found in the spinal cord. The neurotransmitter is formed from glutamic acid by the action of glutamate decarboxylase. Under the action of GABA-transaminase, it is degraded to succinic semialdehyde (Figure 3.4).

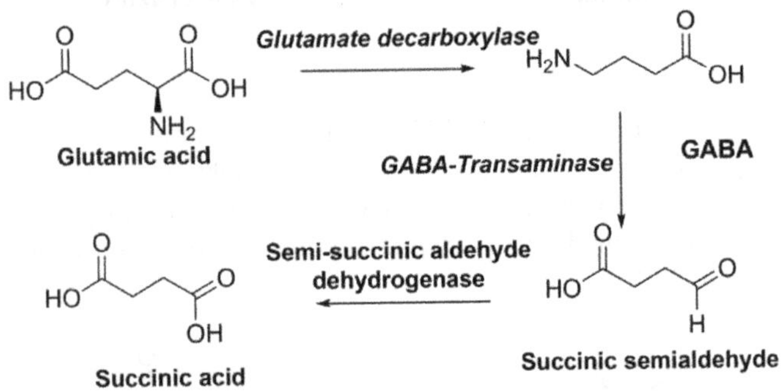

Figure 3.4. *GABA biosynthesis and catabolism*

GABA is involved in the control of neural activity in the cerebral cortex. Deficient GABA concentration in the neurons of the central nervous system is an important factor in the onset of seizure disorders in epilepsy. These disorders are also attributed to an increase in excitatory amino acids such as aspartate and glutamate. Under these conditions, the treatment of this disease involves drugs with different mechanisms of action, with the following objectives:

– potentiate GABAergic activity (GABA agonists), which restores and stimulates inhibitory neurotransmission in the brain, using phenobarbitals, benzodiazepines, *Vigabatrin*, *Gabapentin*, *Felbamate*;

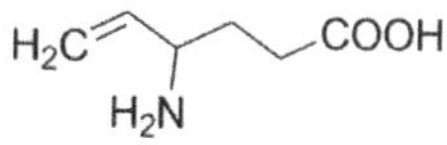

Vigabatrin

– inhibit gamma-amino butyrate transaminase, preventing the degradation of GABA to succinic semialdehyde and increasing GABA concentration in the central nervous system;

– inhibit voltage-dependent sodium channels stabilizing the presynaptic neural membrane, and prevent the release of glutamate and aspartate implicated in the onset, maintenance and propagation of epilepsy.

Some compounds combine these mechanisms of action: *valproic acid*, *Topiramate*. Others follow an unclear mechanism (*Levetiracetam*).

Vigabatrin is a selective inhibitor of this enzyme. It acts as an anti-epileptic. By inhibiting the action of this enzyme, GABA degradation is prevented, and its inhibitory action at the brain level is prolonged. Valproic acid (and its sodium salt), corresponding to dipropylacetic acid, is an inhibitor of this enzyme. It thus exerts an indirect GABAergic action. Valproic acid and its derivatives inhibit folic acid activity. These drugs are used under supervision for the treatment of generalized epilepsy, grand mal seizures, absence seizures, the prevention of hyperthermic convulsions and behavioral disorders associated with epilepsy. They are contraindicated in combination with Mefloquine (risk of convulsions), as well as in cases of liver disease or a history of hepatitis. They should not be prescribed to women of childbearing age or to pregnant women, except in cases of ineffectiveness or intolerance to drug alternatives, due to a high risk of congenital malformations: an average of up to 10% of neural tube defects in the central nervous system of the embryo, cases of spina bifida and cognitive disorders in the newborn.

3.1.3. *Metabolism and endocrinology*

3.1.3.1. *Hydroxyphenylpyruvate dioxygenase (EC 1)*

This enzyme is an iron (Fe^{2+})-dependent enzyme. It has no heme nucleus. It catalyzes the transformation of 4-hydroxyphenylpyruvate into homogentisate. This is the second biochemical reaction involved in tyrosine degradation.

In *tyrosinemia type I*, an autosomal recessive inherited disorder, there is a deficiency of fumarylacetoacetate hydrolase in the catabolic tyrosine pathway. This enzyme is the fourth in the tyrosine degradation pathway.

Nitisinone is a competitive inhibitor of 4-hydroxy-phenylpyruvate dioxygenase, acting on upstream tyrosine catabolism in patients with tyrosinemia type I.

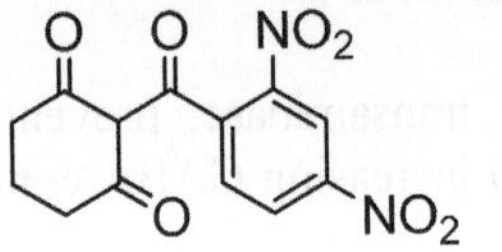

Nitisinone

This molecule prevents the accumulation of toxic intermediates. A diet that controls tyrosine and phenylalanine intake is part of the treatment.

3.1.3.2. *Phenylalanine hydroxylase (EC 1)*

This enzyme, working with the cofactor tetrahydrobiopterin (BH4), takes part in the degradation of excess phenylalanine by converting this amino acid into tyrosine.

Phenylketonuria is a hereditary disease resulting in a deficiency of this enzyme. Phenotypes can range from a very slight increase in blood phenylalanine concentrations to a more severe phenotype, with marked hyperphenylalaninemia. If left untreated, phenylalanine accumulation leads to profound and irreversible mental disability. In mild cases, the administration of *Sapropterin*, or tetrahydrobiopterin, which activates phenylalanine hydroxylase, combined with diet, helps to stabilize the disease.

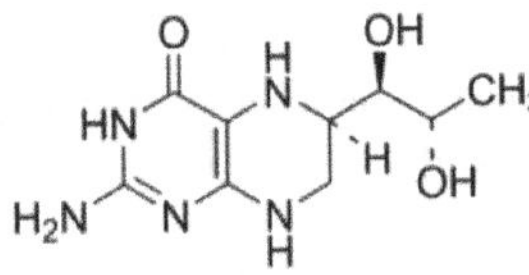

Sapropterin

3.1.3.3. *Beta-hydroxysteroid dehydrogenase (EC 1)*

This enzyme is present in the body in various isoforms. It catalyzes the transformation of pregnenolone into progesterone, 17-hydroxypregnenolone into 17-hydroxyprogesterone, dehydroepiandrosterone into androstenedione and androstenediol into testosterone. Deficiency of this enzyme results in increased androgenic action in women, reduced action in men, and reduced mineralocorticoid and glucocorticoid action. Inhibiting this enzyme may be of interest in syndromes where cortisol is produced in excess, such as Cushing's syndrome.

Trilostane was a drug used to inhibit this human enzyme.

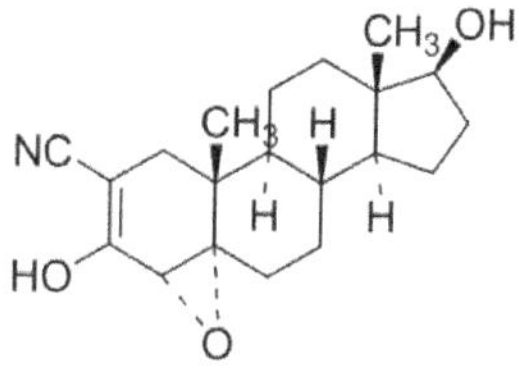

Trilostane

It is currently used in veterinary medicine to treat Cushing's disease in dogs (hypercortisolism). Molecules are currently being researched as enzyme inhibitors. Currently, in humans, this disease is treated by surgery. When surgery is not applicable, combinations of different drugs are available to control excess cortisol in the majority of patients.

3.1.3.4. *Alcohol dehydrogenase (EC 1)*

This dimeric enzyme, which is mainly found in the liver, as well as in the kidney and gastrointestinal tract, exists in several isoforms linked to different monomer-encoding genes. Thanks to this catalyst, primary alcohols are converted into aldehydes in the presence of NAD^+. Isoforms are more specific for ethanol, which they convert into acetaldehyde (ethanal).

Fomepizole is a competitive inhibitor of alcohol dehydrogenase. This drug is used in cases of intoxication by ethylene glycol (an antifreeze compound used as a solvent). It prevents the enzyme from forming the toxic metabolites formed from these two products.

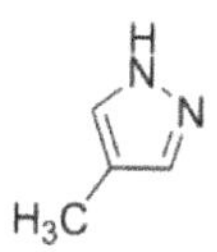

Fomepizole

3.1.3.5. *Aldehyde dehydrogenase (EC 1)*

This enzyme, which exists in several isoforms, catalyzes the conversion of an aldehyde into a carboxylic acid. In the liver, ADH II, one of several human isoforms of the enzyme, converts ethanal into acetic acid.

Enzyme inhibition by *Disulfiram*, used in the treatment of alcohol dependence, works by preventing the transformation of glyceraldehyde formed from alcohol into acetic acid. The resulting accumulation of glyceraldehyde leads to alcohol intolerance. This intolerance leads the alcohol-dependent individual to avoid alcohol.

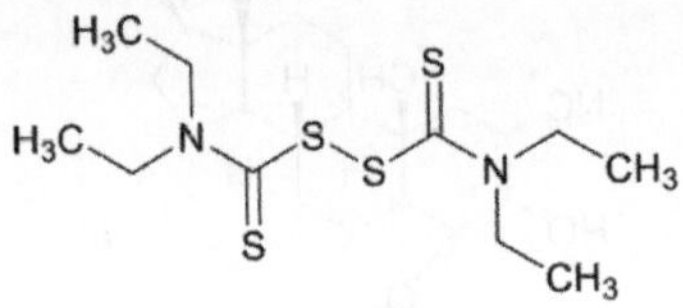

Disulfiram

3.1.3.6. *Dipeptidyl peptidase (several isoenzymes) (EC 3)*

Dipeptidyl peptidases (DPPs) are catalysts for the cleavage of numerous physiological peptides. When dysfunctional, these enzymes are associated with a wide range of pathologies, including Crohn's disease, chronic liver disease, osteoporosis, many scleroses, eating disorders, rheumatoid arthritis, cancer and type II diabetes.

Diabetes mellitus is a growing disease, affecting over 350 million people according to the WHO. Characterized by higher-than-normal blood glucose levels (hyperglycemia), diagnosis is established by determining blood glucose levels and glycated hemoglobin concentration (HbA1c). In healthy subjects, fasting blood glucose must be less than 1.10 g/l and less than 1.40 g/l at 120 minutes from the start of an induced hyperglycemia test.

Homeostasis of blood glucose is essentially ensured by two pancreatic hormones:

– insulin, a hypoglycemic hormone secreted by the beta cells of the islets of Langerhans in the pancreas, which enables muscle tissue and the liver to store glucose in the form of glycogen;

– glucagon, a hyperglycemic hormone secreted by pancreatic alpha cells, releases glucose from hepatic glycogen.

Subsequent discoveries have completed this picture, involving other mechanisms for regulating blood glucose levels. After food intake, peptide hormones (incretins) are secreted by cells in the gastrointestinal tract (Arnoff et al. 2004):

– GIP (glucose-dependent insulinotropic polypeptide), secreted in the duodenum and jejunum;

– GLP-1 (glucagon-like peptide-1), a hormone secreted in the small intestine and ascending colon.

GLP-1 has various biological activities: it stimulates glucose-dependent insulin secretion (incretin effect) by acting on G-protein-coupled receptors (GPCRs) in the pancreas. It suppresses the secretion of hyperglycemic glucagon, which reduces hepatic glucose production. It slows gastric emptying and food intake. It is also involved in the survival and proliferation of pancreatic beta cells. Its use is of great interest in lowering blood glucose levels in diabetics.

The structure of the 37-amino-acid human GLP-1 (hGLP-1) peptide comprises various biologically active isoforms: an isoform that contains amino acids (7-37), an isoform that presents amino acids (7-36) and a shortened GLP-1 isoform (9-36) which no longer contains the terminal amino acid (Box 3.1).

7
H_2N-His-Ala-Glu-Gly-Thr-Phe-Thr-Ser-Asp-Val-Ser-Ser-Tyr-Leu-Glu-Gly-Gln-Ala-Ala-
Lys-Glu-Phe-Ile-Ala-Trp-Leu-Val-Lys-Gly-Arg-Gly-COOH
26 34 36 37

Box 3.1. *GLP-1 peptide sequence*

The peptide structure of GLP-1 makes it unsuitable for oral use, as it is degraded by dipeptidylpeptidase IV (DPP-4). DPP-4 leads to the cleavage of a His(7)-Ala(8) dipeptide fragment, causing loss of activity of the peptide. In order to circumvent this degradation and benefit from GLP-1's favorable glycemic control effects, two pathways are implemented:

– Prepare more stable GLP-1 analogues resulting from the substitution of certain amino acids. This approach has made it possible to introduce these derivatives into therapeutics. There are currently short-acting GLP-1 receptor agonists requiring one to two doses per day, and derivatives requiring weekly administration.

– DPP-4 inhibitors, to prevent the degradation of GLP-1 and GIP released after food intake, so as to maintain GLP-1 activity.

The dipeptidylpeptidase IV (DPP-4) inhibitors currently available are: *Vildagliptin*, *Saxagliptin*, *Sitagliptin*, *Linagliptin* (Figure 3.5). By preventing the degradation of GLP-1 and GIP released after food intake, these DPP-4 inhibitors increase the blood concentration of both peptides, as well as the benefits of their effects on glycemia control. This represents a new approach to the treatment of diabetes.

Various GLP-1 analogues have been introduced into therapeutics: *Exenatide*, *Liraglutide*, *Lixisenatide*, *Albiglutide*, *Dulaglutide* and sustained-release *Exenatide*. GLP-1 receptor agonists have a direct action on the feeling of satiety, whereas DPP-IV analogues do not.

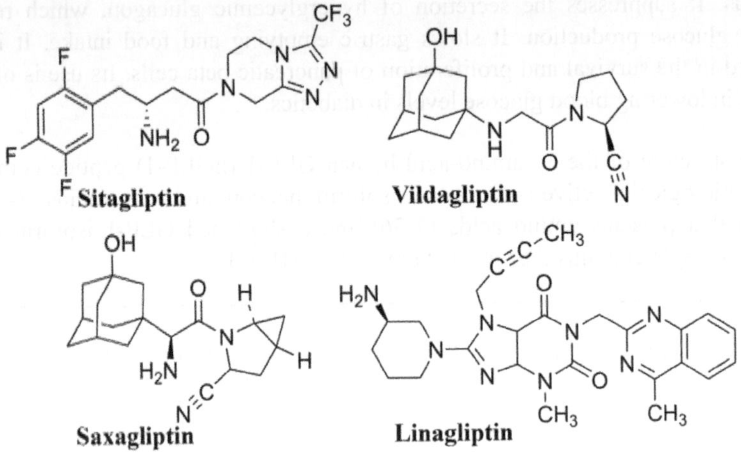

Figure 3.5. *Dipeptidylpeptidase IV inhibitors*

Various combinations of DPP-4 inhibitors have also been introduced into the diabetes arsenal as supplements to insulin or Metformin: Sitagliptin + Metformin and Vildagliptin + Metformin. Other associations are also available.

A great deal of research is being carried out in various fields on the various DPP isoenzymes.

3.1.3.7. *Alpha-glucosidases (EC 3)*

Dietary carbohydrates are absorbed directly by the intestinal mucosa in the case of monosaccharides, and after hydrolysis by alpha-glucosidases in the case of disaccharides, oligosaccharides and polysaccharides. Absorption of sucrose and starch from the intestinal villi takes place in the form of glucose units after hydrolysis by alpha-glucosidases. Reversible, specific inhibition of alpha-glucosidases (saccharase, maltase, amylase, dextrinase) slows down carbohydrate hydrolysis in the glucosidase-rich upper part of the small intestine (duodenum) and shifts it to the ileum and colon, thereby reducing glucose absorption. It thus prevents the postprandial hyperglycemic spike accompanied by hyperinsulinism and glycosuria.

Acarbose is an alpha-glucosidase inhibitor. Its structure corresponds to a pseudotetrasaccharide with an acarviosine unit linked to maltose. It is obtained by extraction from *Actinoplanes utahensis* bacterial culture broths (Figure 3.6). The acarviosine fragment of acarbose has a higher binding affinity to the active site of alpha-glucosidases than the normal substrates of these enzymes (sucrose, maltose, starch, dextrins). These properties are used to treat non-insulin-dependent diabetes.

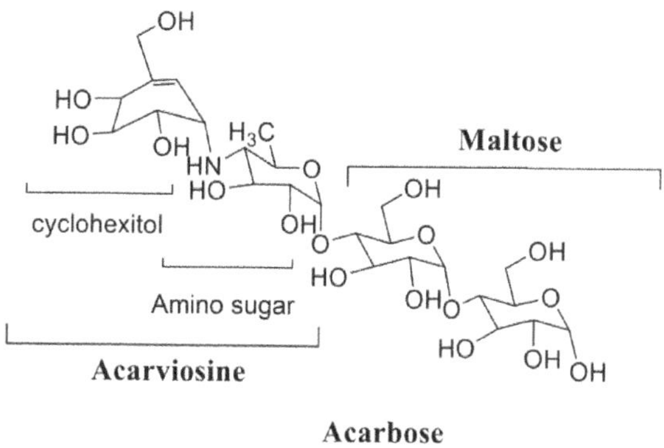

Figure 3.6. *Acarbose structure*

Miglitol (Figure 3.7) is a synthetic product whose structure corresponds to a substituted piperidine, hexose analogue, or hydroxyethyl-deoxynojirimycin (Figure 3.7). It has a specific, reversible inhibitory effect on intestinal alpha-glucosidases, similar to that of Acarbose.

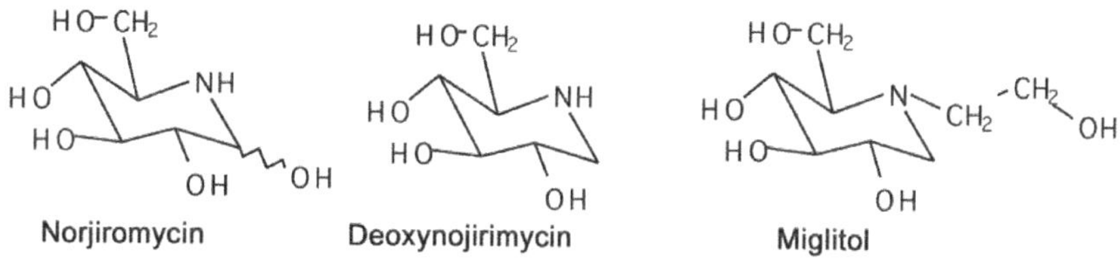

Figure 3.7. *Structure of Norjiromycin, Deoxynojirimycin and Miglitol*

Miglitol is also used in the treatment of non-insulin-dependent type II diabetes, slowing carbohydrate absorption and preventing hyperglycemia. Alpha-glucosidase inhibitors are additional treatments for non-insulin-dependent type II diabetes. These include the various therapeutic families: sulfonylureas, glinides, biguanides, thiazolidinediones, sodium-glucose cotransporter 2 (SGLT-2) inhibitors, GLP-1 analogues and dipeptidyl peptidase 4 (DPP-4) inhibitors.

3.1.3.8. *ATPase H^+/K^+ (EC 3)*

This is a membrane enzyme known as the "proton pump". This pump is activated by gastrin and histamine.

At the gastric level, ATP hydrolysis by the pump provides the energy needed to drive protons out into the gastric lumen and potassium (K^+) ions into gastrocytes. This is active antiporter transport. The outflow of protons will acidify the gastric juice. Prolonged, excessive acidity of this juice leads to ulcers.

The discovery of proton pump or H^+/K^+ ATPase enzyme inhibitors followed the development of the anti-acid properties of *Timoprazole*, whose toxicity did not lead to its therapeutic use. Structural modifications subsequently led to *Omeprazole* for the treatment of duodenal ulcers and gastro-esophageal reflux disease.

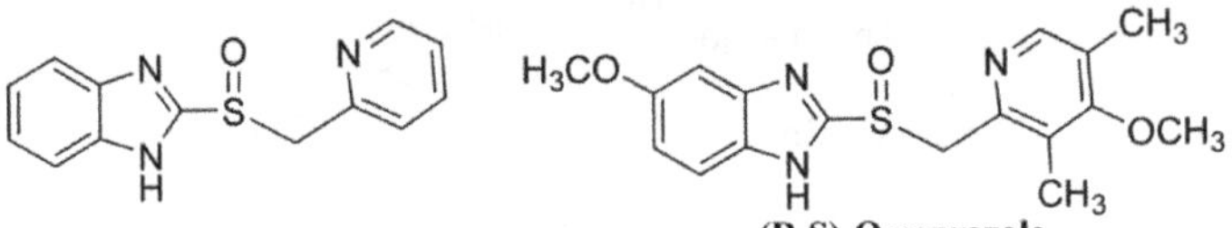

Timoprazole **(R,S)-Omeprazole**

They are taken orally, and once taken, thanks to a gastro-resistant coating, the proton pump inhibitors, which are unstable in an acid environment, pass through the stomach without being degraded. Once they reach the intestinal tract, they are released and absorbed by the gastric parietal cell. They selectively inhibit the H^+/K^+ ATPase or "proton pump" enzyme. They thus prevent the release of H^+ ions, leading to a drop in gastric acidity. Stomach pH rises to approximately 4–5.

In the case of precursors, their action is the consequence of their transformation into an activated cyclic form (sulfenamide). These cyclic forms bind covalently to the extracytoplasmic loop of the proton pump, at the level of cysteine residues, and thus block the transport of H^+ ions to the stomach. Ulcer cicatrization is achieved after a few weeks of treatment.

Subsequent work showed that the S-enantiomer is more active than the R-derivative, and was introduced under the *Esomeprazole* CID.

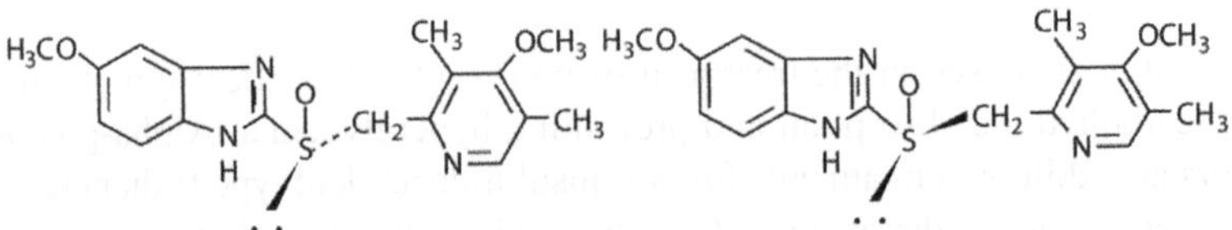

S-Omeprazole: Esomeprazole **R-Omeprazole**

New compounds inhibiting this enzyme were subsequently developed (Figure 3.8). Their structure comprises a substituted benzimidazole heterocycle with anti-acid activity, a sulfoxide function and a pyridine ring, variously substituted depending on the compound.

Derivatives marketed include *Omeprazole*, *Lansoprazole*, *Pantoprazole* and *Rabeprazole*. Clinical trials are underway to develop new derivatives.

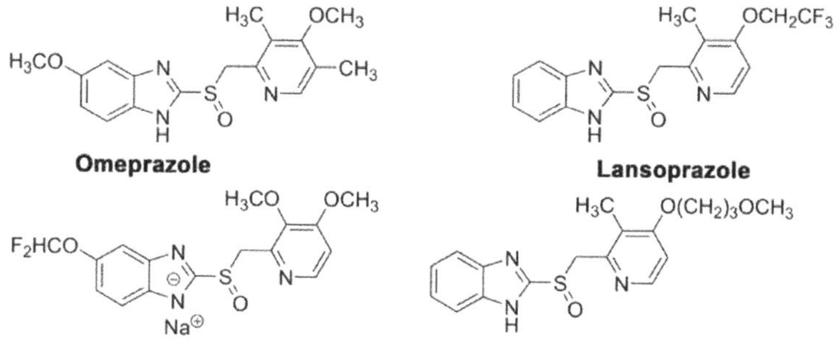

Figure 3.8. *H^+/K^+ ATPase pump inhibitors*

3.1.3.9. *HMG-CoA reductase (EC 1)*

This enzyme catalyzes the conversion of (beta)-hydroxy-(beta)-methyl-glutaryl-CoA (HMG-CoA) to mevalonate, in the presence of NADPH. This is a key enzyme in cholesterol biosynthesis. This enzyme at the start of the metabolic pathway is inhibited by cholesterol, the end product of the pathway. Phosphorylation of this enzyme by a kinase (EC 2) inactivates the enzyme.

Plasma cholesterol levels depend on dietary intake and the various stages of biosynthesis (Figure 3.9). It is the raw material for the biosynthesis of sex hormones, mineralocorticoids, glucocorticoids, vitamin D and bile acids.

3-Hydroxy-3-methyl-glutaryl-coenzyme-A reductase is an enzyme involved in biosynthesis at the mevalonate stage (Figure 3.9). Inhibitors of this enzyme prevent the conversion of 3-hydroxy-3-methylglutaryl-coenzyme A into mevalonic acid, the precursor of sterols and cholesterol.

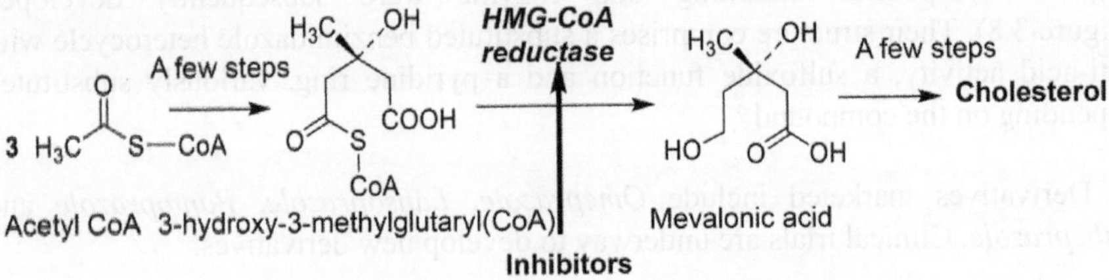

Figure 3.9. *Cholesterol biosynthesis and site of action of HMG-CoA reductase inhibitors*

Biologically speaking, cholesterol and triglycerides that have been biosynthesized in the liver are incorporated into VLDLs (very low-density lipoproteins) and released into plasma, from where they reach peripheral tissues. VLDLs enable the formation of LDLs (low-density lipoproteins), which are recognized at high-affinity LDL receptors and subsequently catabolized.

By inhibiting HMG-CoA reductase, these drugs lower hepatic cholesterol biosynthesis. They induce the overexpression of the LDL receptor gene, increasing the synthesis of LDL receptor messenger RNA, which in turn increases the number of receptors in hepatocytes. The result is increased LDL uptake and catabolism.

HMG-CoA reductase inhibitors can be divided into two groups:

– The first products are of natural origin, derived from fungal metabolism or hemisynthesis. These include *Mevastatin* (*Penicillium citrinum*), whose toxicity has not led to its being marketed. *Lovastatin* (*Aspergillus*) and *Simvastatin* are precursors of *Pravastatin*, the active acid form of the drug (Figure 3.10).

– Derivatives prepared by total synthesis include: *Fluvastatin*, *Atorvastatin*, *Cerivastatin* and *Pitavastatin*. Cerivastatin has been withdrawn since 2001, due to the occurrence of severe rhabdomyolysis (skeletal muscle damage) (Figure 3.11). The appearance of cases of rhabdomyolysis in patients treated with this molecule occurs when it is combined with Gemfibrozil, a hypolipidemic (Ozdemir et al. 2000). Muscular damage occurs, with a sharp rise in creatine phosphokinase (CPK) levels, myoglobinemia and myoglobinuria, and may even lead to renal failure.

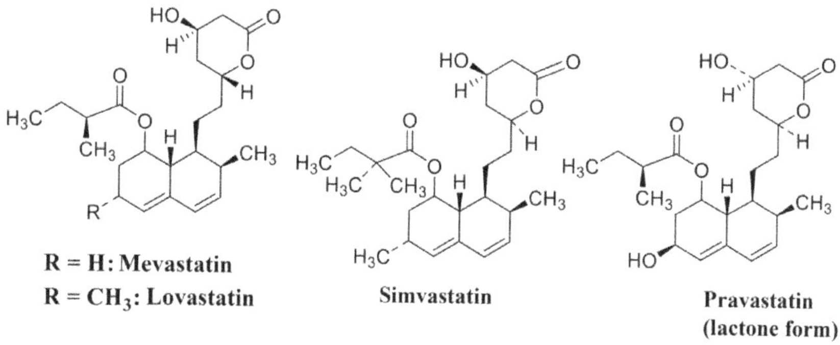

Figure 3.10. *Natural and hemisynthesis inhibitors of HMG-CoA reductase*

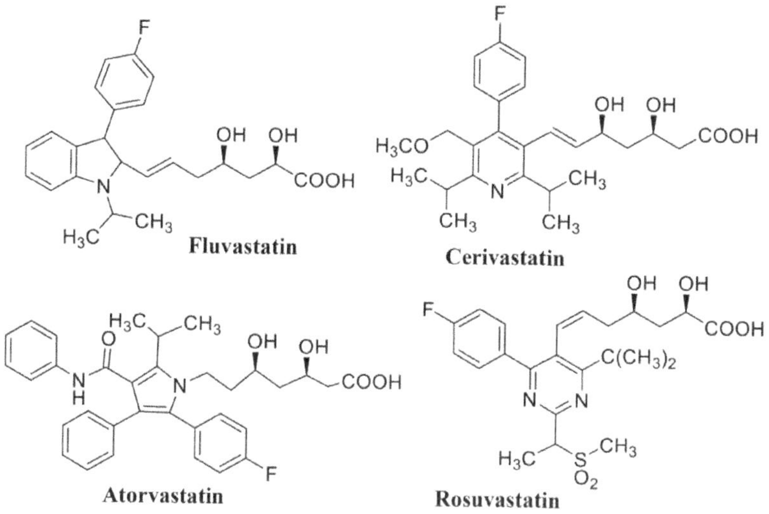

Figure 3.11. *HMG-CoA reductase synthesis inhibitors*

Studies on the relationship between the structure and activity of Pravastatin, Fluvastatin, Atorvastatin and Pitavastatin show that the free carboxylic acid function enables them to be directly active. On the other hand, derivatives with a lactonic structure are precursors leading to the active forms, after opening and formation of the corresponding dihydroxylated acids (Figure 3.12), (3R,5R)-3,5-dihydroxyheptanoic acid.

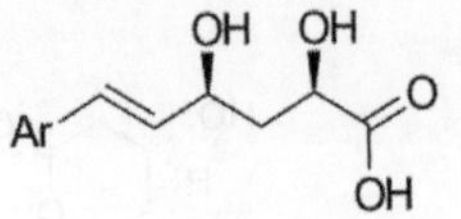

Acid structure: (3R, 5R)-3,5-dihydroxyhept-6-enoic acid active form

Figure 3.12. *Active forms of HMG-CoA reductase inhibitors*

On the contrary, acetylation or blocking of the hydroxyl functions, or inversion of their configuration, lowers activity. The hexahydronaphthalene ring is important, and determines the enzyme's binding affinity. In synthetic derivatives, it is replaced by an indole (Fluvastatin) or pyrrole (Atorvastatin) heterocycle.

These derivatives reduce the hepatic biosynthesis of cholesterol and VLDL-cholesterol. They are used to treat hyper-cholesterolemia and hyper-triglyceridemia. They are also prescribed for primary prevention in patients with cardiovascular risk factors, when hygienic-dietary measures are not sufficient to reduce cholesterol levels. In secondary prevention, they are useful for patients with a history of myocardial infarction, coronary insufficiency or a stroke.

Rosuvastatin has the same indications as other statins.

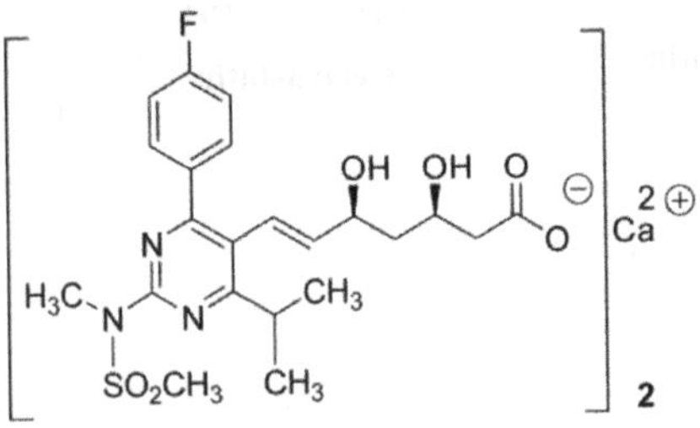

Rosuvastatin

Statins have the disadvantage of causing muscle damage and should not be combined with fibrates or *Amiodarone*.

Various studies report an increased risk of diabetes when using certain statins compared to Pravastatin, particularly in the case of Rosuvastatin, Atorvastatin and Simvastatin (Carswell et al. 2002). Risks of inflammatory necrotizing myopathy are also possible for statins (Padala and Thompson 2012; Carter et al. 2013).

3.1.3.10. *Triacylglycerol lipases (pancreatic and gastric) (EC 3)*

Lipases hydrolyze triglycerides into fatty acids and monoglycerides, which are then absorbed in the jejunum. Triglycerides are then reconstituted in the enterocyte and combine with various lipid compounds (cholesterol, phospholipids, apolipoproteins) to form large lipoproteins and chylomicrons, which pass into the venous circulation via the lacteal vessels.

After entering the bloodstream, lipoprotein lipases hydrolyze triglycerides, releasing fatty acids which are used as an energy source by muscle cells or stored in adipocytes.

Orlistat, a tetrahydrogen derivative of Lipstatin, secreted by *Streptomyces toxytrini*, inhibits the hydrolytic action of gastric and pancreatic lipases on triglycerides, preventing the release of fatty acids and monoglycerides. This action enables triglycerides to pass through the gastrointestinal tract without being absorbed, facilitating their removal through feces. This action, which is the result of a covalent bond of this "decoy" substrate with the active site of lipases, is responsible for its lipid-lowering activity (Chen et al. 2017) (Figure 3.13).

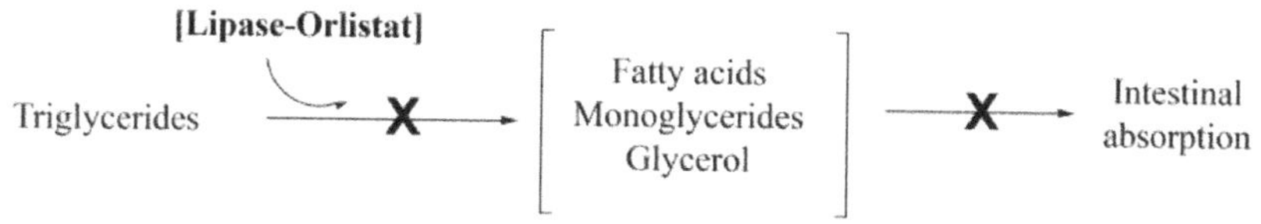

Figure 3.13. *Mechanism of action of Orlistat*

3.1.3.11. *Xanthine oxidase (EC 1)*

Xanthine oxidase is an enzyme involved in the metabolism of purine bases (adenine, guanine) to form uric acid (Figure 3.14).

This enzyme is involved in the metabolism of uric acid, an excess of which is responsible for gout. Gout can be treated with various hypouricemic agents that inhibit its synthesis, or uricosurics that facilitate urinary excretion (Probenecid, Sulfinpyrazone, Benzbromarone).

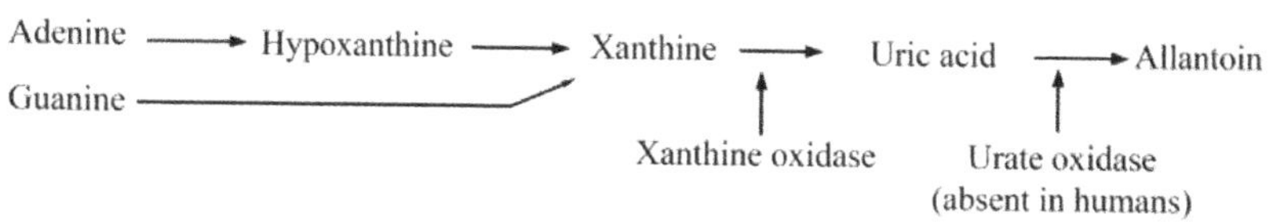

Figure 3.14. *Purine base catabolism*

Allopurinol, *Tisopurine* and *Febuxostat* inhibit xanthine oxidase and therefore uric acid biosynthesis (Figure 3.15). The increase in the level of undegraded purine nucleotides acts as a feedback inhibitor of amidophosphoribosyl-transferase (EC 2), an enzyme involved in purine biosynthesis, and therefore indirectly in uric acid biosynthesis.

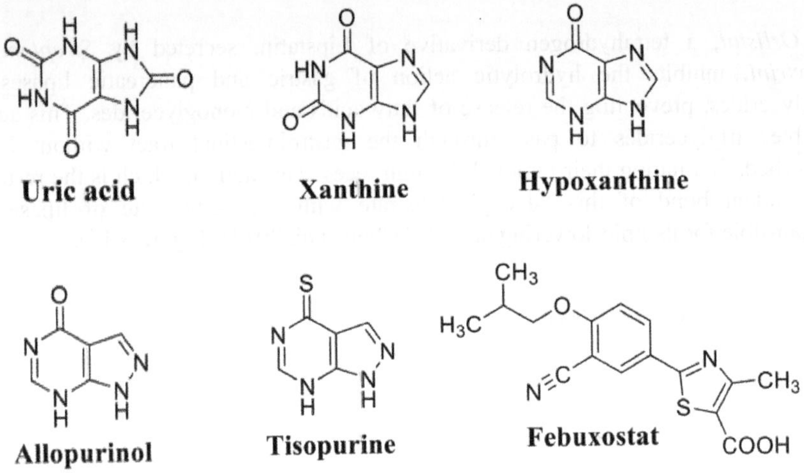

Figure 3.15. *Xanthine oxidase inhibitors*

3.1.3.12. *Enkephalinase (EC 3)*

This catalyst hydrolyzes exogenous and endogenous peptides, in particular enkephalins, hence the name. It is a membrane-bound enzyme present in a variety of tissues. In the digestive tract, enkephalins have an antisecretory action, especially in the intestine.

Opioid peptides are a group of neuropeptides that includes enkephalins, endorphins and dynorphins. They are formed from a precursor polypeptide for each molecule in the same family.

Enkephalins are found in various structures of the central nervous system (spinal cord, limbic structures, striatum, raphe nucleus, hypothalamus, etc.) and exert their pain-modulating effect by favorably binding to morphine receptors.

Their structures were established through the work of Hughes and Kosterlitz (Kosterlitz and Hughes 1977):

Methionine enkephalin: H_2N – Tyr – Gly – Gly – Phe – Met – OH

Leucine enkephalin: H_2N – Tyr – Gly – Gly – Phe – Leu – OH

Beta-endorphin (Li and Smyth (Smyth 2016)) is a 31-amino acid peptide with potent analgesic activity. Its terminal portion is that of methionine-enkephalin. The 31-amino acid sequence of beta-endorphin is identical to amino acids 60–91 of pituitary β-lipotropin. Beta-endorphin activates μ and δ receptors, but has a low binding affinity for κ receptors.

In addition to their central and peripheral antinociceptive functions, opioid peptides can modulate immune activity and cell proliferation. Enkephalins and endorphins are orally inactive. They have a short duration of action, as they are inactivated by enkephalinase hydrolyzing the Gly–Phe bond. Modified enkephalins have been developed to produce stable, orally usable products. In order to prolong their antinociceptive action, natural L-series amino acids were replaced by D-series amino acids. These experiments did not lead to conclusive results. In contrast, peptide screening revealed enkephalinase inhibitory activity by the dipeptide phenylalanyl-glycine. Replacement of the dipeptide's amine function by a thioacetyl group forms Racecatodril (precursor), generating the active metabolite with a thiol function (Figure 3.16). Racecatodril is an enkephalinase inhibitor used as an antidiarrheal agent for the symptomatic treatment of acute diarrhea (Llorens et al. 1980).

Current research is continuing into the development of new analgesic enkephalinase inhibitors.

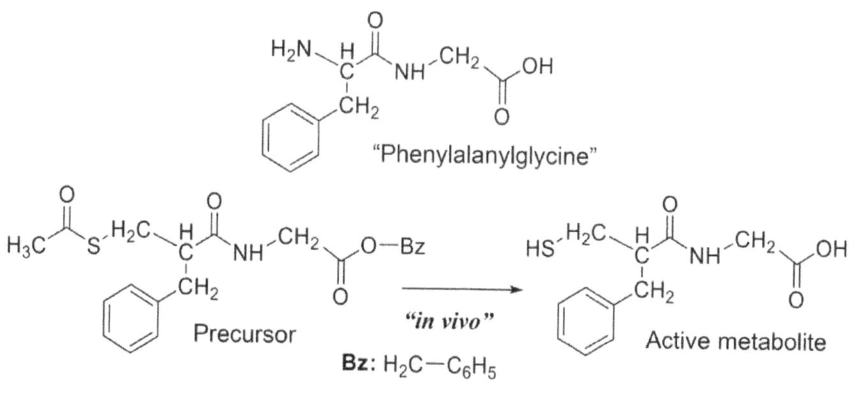

Figure 3.16. *Racecatodril metabolism*

3.1.4. *Cardiovascular and immunology*

3.1.4.1. *Phosphodiesterases (PDE–EC 3)*

cGMP phosphodiesterases catabolize cGMP, the second messenger, when its biological action is complete. cAMP phosphodiesterases do the same with cAMP, another secondary messenger in signaling pathways. There are several isoforms of these enzymes. Some are specific for cGMP (PDE5, PDE6, PDE9), others for cAMP (PDE4, PDE7, PDE8), and the last ones recognize both (PDE1, PDE2, PDE3, PDE10 and PDE11) (Conti and Beavo 2007; Francis et al. 2011).

The enzyme catalyzes a hydrolysis reaction of the ester bond between the alcohol function carried by the 3' carbon of the ribose and the phosphate carried by the 5' carbon of the cyclic nucleotides. This phosphoric ester bond is at the origin of the cycle that makes the molecule active. Its hydrolysis produces a 5' phosphate nucleotide that is no longer cyclic, and therefore inactive.

PDEs are the target of various therapeutic applications (Keravis and Lugnier 2012).

Numerous inhibitors, either validated or under research, target these isoenzymes:

– Xanthine, theophylline and caffeine derivatives exhibit diuretic and bronchodilator properties (Schudt et al. 2011; Barnes 2013). They are non-selective PDE inhibitors that increase cAMP concentration. As this inhibition is not selective of type 3A phosphodiesterase, it leads to significant side effects through inhibition of other types of PDE. The same applies to Dipyridamole used as an antiplatelet agent (Balakumar et al. 2014).

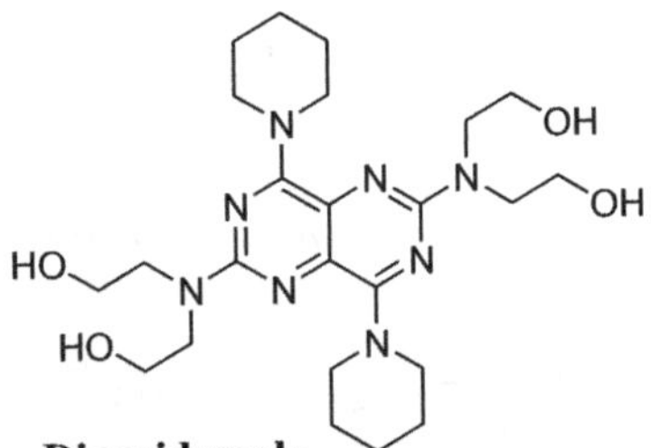

Dipyridamole

The development of selective PDE 3A inhibitors should make it possible to develop anti-asthmatics and anti-inflammatories that do not induce adverse effects that can be observed with theophylline.

– *Anagrelide* is a cAMP and cGMP phosphodiesterase 3 inhibitor.

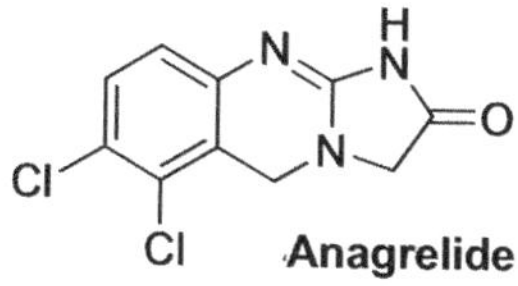

Anagrelide

This molecule reduces the number of platelets in thrombocythemia. The drug acts by inhibiting this phosphodiesterase. In vitro and in vivo studies have shown that Anagrelide acts selectively on platelets.

– *Amrinone* and Milrinone also act as phosphodiesterase 3 inhibitors in the heart.

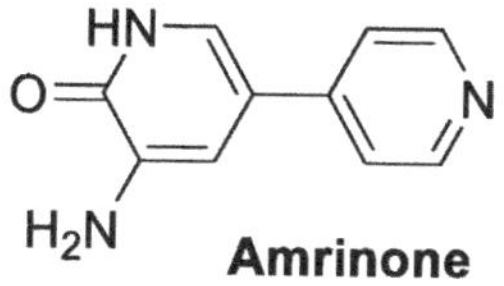

Amrinone

The latter two drugs are positive ionotropic agents (influx of calcium into cells) with vasodilatory properties. During treatment, it is important to monitor the possible occurrence of iatrogenic effects, particularly at the platelet level.

– *Sildenafil*, *Vardenafil* and *Tadalafil* are treatments that improve erectile function in men by acting on PDE-5.

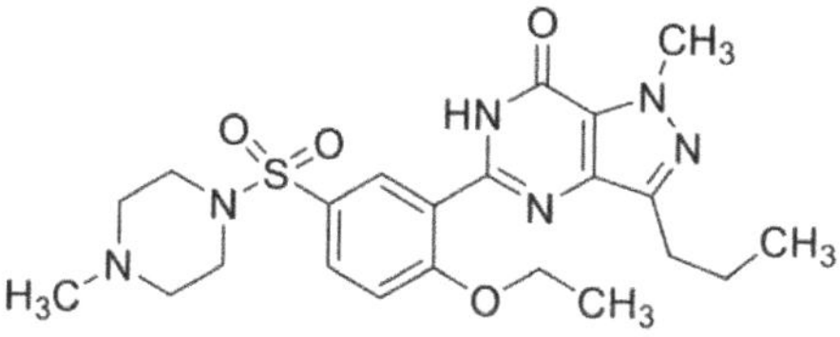

Sildenafil

Erectile dysfunction is restored by increasing blood flow to the penis. During sexual activity, nitric oxide released in the corpora cavernosa enables erection by stimulating the production of cGMP via guanylate cyclase (EC 4). cGMP causes smooth muscle relaxation and, consequently, blood flow. These molecules inhibit cGMP phosphodiesterase 5, thereby increasing cGMP half-life and erection duration. Numerous inhibitors of this type are described in the literature (Corbin et al. 2002).

– PDE4 inhibitors have strong therapeutic potential as anti-inflammatories and memory-enhancing derivatives (Zhang et al. 2005).

3.1.4.2. *Cyclooxygenases (COX–EC 1)*

Prostaglandin (PG) biosynthesis involves the initial intervention of Phospholipase A2 to form a fatty acid (arachidonic acid), followed by the cyclooxygenase pathway leading to the various PGs, or lipoxygenase to form the other prostanoids, including leukotrienes (Figure 3.17).

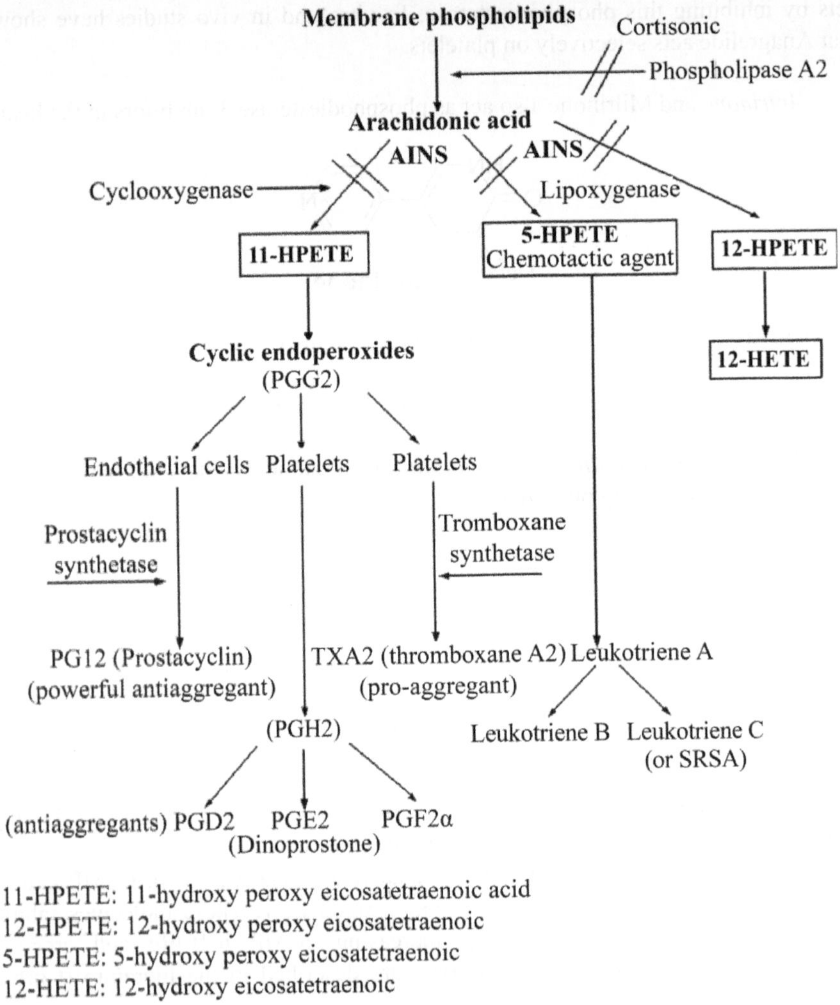

Figure 3.17. *Biosynthesis of prostaglandins and leukotrienes*

PGs are remarkable for their multiple actions, exerted through cyclic nucleosides. PGs formed by the lipoxygenase (lipoxidase) pathway have activity on the vascular system, and those formed from cyclooxygenase (COX) have activity on cell migration and inflammatory reactions:

– Algogenic action: by sensitizing vascular nerve endings to bradykinin and histamine.

– Phlogogenic action: through a chemotactic effect. PGs activate the migration of leukocytes and monocytes from vessels into the exudate of the inflamed joint, leading to the production of lysosomal enzymes and edema.

– Leukotrienes are involved in immune, allergic and inflammatory reactions. Leukotriene C (SRSA: Slow Reacting Substance of Anaphylaxis) is a bronchospasm factor also involved in inflammatory and osteoarthritic reactions.

Free radicals formed are also involved in situ in inflammatory reactions, as well as PAF (platelet activating factor), interleukins, in particular interleukin-1gamma (a pro-inflammatory cytokine), tumor necrosis factor (TNFalpha), a pro-inflammatory cytokine that activates T lymphocytes and fibroblast proliferation, and various metalloproteinases (collagenases and proteoglycanases). These enzymes are responsible for the destruction of cartilage (chondrolysis) and joints.

Interleukin-1 induces inflammatory activity through a variety of actions: PGE2 synthesis, protease production, bone catabolism (action of osteoclasts), cartilage degradation (enzyme secretion by chondrocytes) and fibroblast stimulation. It also contributes to the onset of fever through direct action on the hypothalamus. It facilitates the diapedesis of monocytes and granulocytes, releasing their mediators at the inflammation site.

Natural prostaglandins (PGE1, PGE2) exert a protective effect on the gastric mucosa, by reducing the basal acid flow and increasing mucus and bicarbonate secretion.

Since 1989, two cyclooxygenases (COX) encoded by two different genes have been known:

– The cell's constitutive COX-1 is present in a physiological state in most tissues (digestive tract, kidneys, uterus, platelets, etc.). COX-1-derived PGs have a protective action on the gastric mucosa. Their inhibition by anti-inflammatory drugs is responsible for adverse effects such as prolonged bleeding time, damage to the gastric mucosa and reduced glomerular filtration.

– Inducible COX-2 is not expressed in normal tissues. It is produced in response to inflammatory stimulation by cytokines and polysaccharides. PGs produced by

COX-2 have a pro-inflammatory action. Its inhibition leads to an anti-inflammatory and analgesic action.

Conventional anti-inflammatory agents inhibit both COX-1 and COX-2, which accounts for both their anti-inflammatory action and their toxicity on the gastric mucosa. Selective COX-2 inhibitors, on the other hand, block the inflammatory process and pain without disrupting the gastrointestinal cytoprotective role of PGs, thus preventing undesirable effects on the gastric mucosa and kidney. Under these conditions, research has led to the development of selective COX-2 inhibitors with no effect on COX-1, whose gastrointestinal cytoprotective role is maintained, enabling good tolerance by the gastric mucosa. However, there are more cardiovascular side effects with COX-2 inhibitors than with COX-1 inhibitors.

Non-steroidal anti-inflammatory drugs (NSAIDs) and anti-rheumatic drugs, which are used to treat acute or chronic inflammatory pain, particularly in rheumatic diseases, exert COX inhibitory activity during PG biosynthesis. They can be divided into two groups, according to the selectivity of their action on COX:

– *non-selective NSAIDs*, which act simultaneously on COX-1 and COX-2. Although their structures vary, these derivatives all have one chemical feature in common: they generally have an acid or acid-like function:

- aryl-carboxylic acids (salicylic acid, fenamic acid),
- arylacetic acids,
- 2-arylpropionic acids,
- oxicams (acidity due to the enol group),
- pyrazole derivatives (acidity due to the enol group);

– *selective NSAIDs*, which act selectively on COX-2. Selective NSAIDs act through their anti-inflammatory, analgesic and antipyretic properties, to a greater or lesser extent, depending on the compound.

In addition to COX-inhibiting NSAIDs, osteoarthritis treatment also makes use of monoclonal antibodies and disease-modifying agents. Drugs acting on the cytokines TNFalpha (Tumor Necrosis Factor alpha) have also been developed.

3.1.4.2.1. Non-selective anti-inflammatories inhibiting COX-1 and COX-2

Arylcarboxylic acids and related products

Salicylic acid, and sodium and lithium salts, were once used for their anti-rheumatic properties. Salicylic acid esters (methyl salicylate and phenyl salicylate) are reserved for external use. *Acetylsalicylic acid* has anti-inflammatory, analgesic, antipyretic and antiplatelet-aggregating properties. This action results from the inhibition of COX-1 by acetylation of the hydroxyl groups in the enzyme's active site (transesterification). Another activity is the irreversible inhibition of thromboxane A2 synthesis. Metabolism of acetylsalicylic acid forms salicylic acid by hydrolysis, and gentisic acid by hydroxylation (Figure 3.18). Elimination is mainly carried out in the form of glycuronides, ether sulfate and salicyluric acid.

Various forms of use and therapeutic characteristics are possible. They can be used for long-term treatment of chronic inflammatory rheumatism and arthritis. Short-term treatments are also possible: articular and abarticular rheumatism, arthrosis. They are used in traumatology: fractures, sprains and post-operative pain, as well as in gynecology, stomatology and otorhinolaryngology.

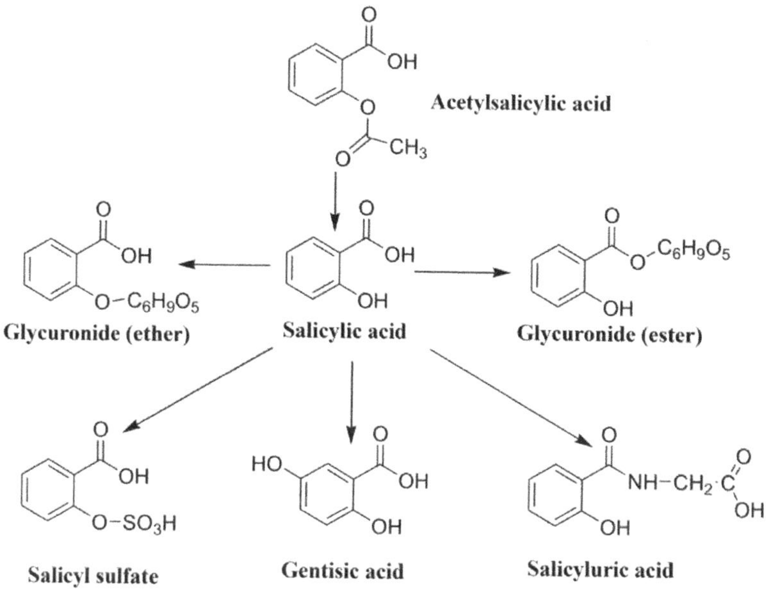

Figure 3.18. *Metabolism of acetylsalicylic acid*

Contraindications include allergy to acetylsalicylic acid and salicylates, ulcers (risk of bleeding), a combination of oral anticoagulants and methotrexate, and third trimester pregnancy. Undesirable effects include digestive disorders, mucosal lesions, acidity, a peptic ulcer, risk of hemorrhage and perforation (requiring the use of a synthetic prostaglandin, Misoprostol, in order to reduce this risk in the case of long-term use. Allergic disorders (pruritus, asthma attacks, anaphylactic shock, bullous dermatoses), renal accidents (oliguria, renal failure) and neuropsychological disorders (headaches, dizziness, insomnia) are also worth noting.

Numerous derivatives with the biological characteristics of acetylsalicylic acid have been developed (Figure 3.19).

Diflunisal has a reversible inhibitory effect on prostaglandin synthesis, with no effect on platelet aggregation. Used as a long-acting analgesic (approximately 10 hours), it can be used to treat rheumatic conditions.

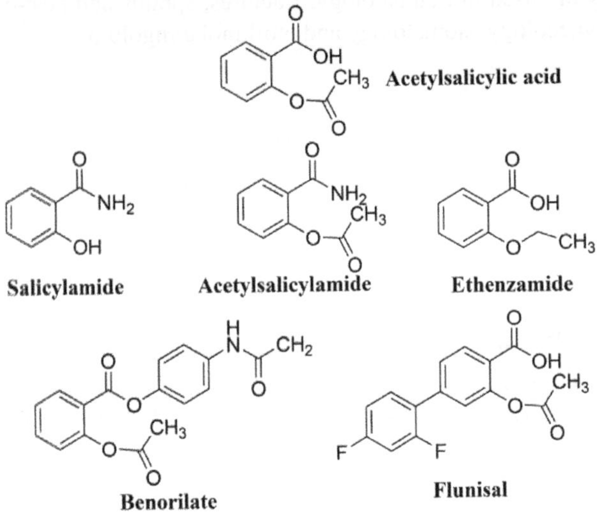

Figure 3.19. *Salicylic and acetylsalicylic acid derivatives*

Anthranilic acid derivatives (fenamic acids) and related products are structurally similar to salicylic acid.

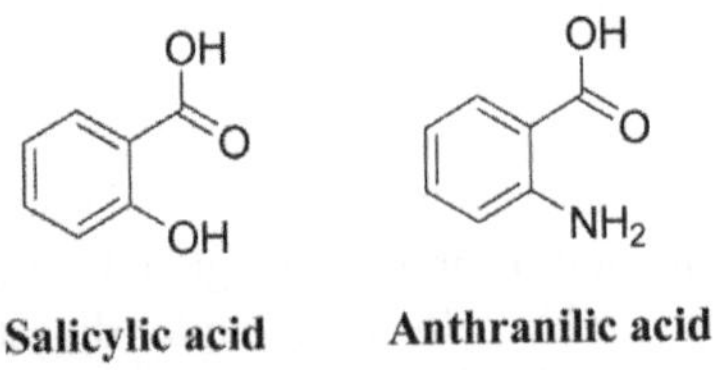

They have both anti-inflammatory and analgesic activities. They have the same indications as other NSAIDs (Figure 3.20).

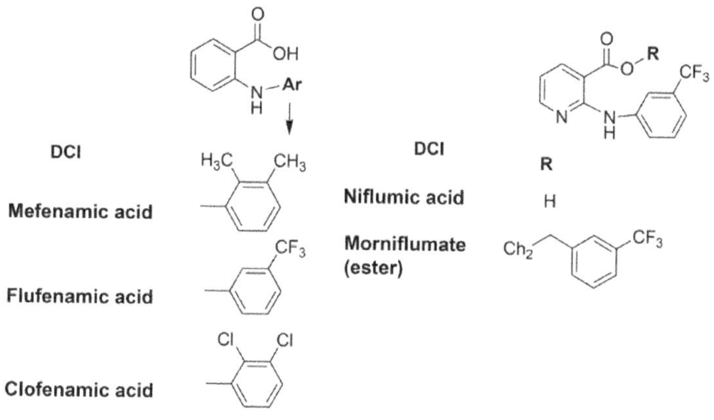

Figure 3.20. *Anthranilic acid derivatives (fenamic acid) and related products*

Other carboxylic derivatives with the same indications as NSAIDs

Ketorolac has a structure with a carboxylic acid function (Muchowski et al. 1985). It is a racemic mixture. It can also be considered similar to acetic and propionic derivatives.

Etodolac is similar in structure to its predecessor and to NSAIDs of the indolacetic series. It is a racemic derivative.

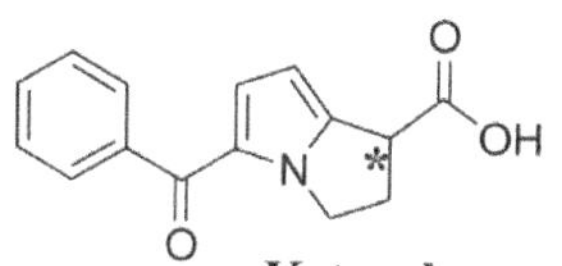

Ketorolac

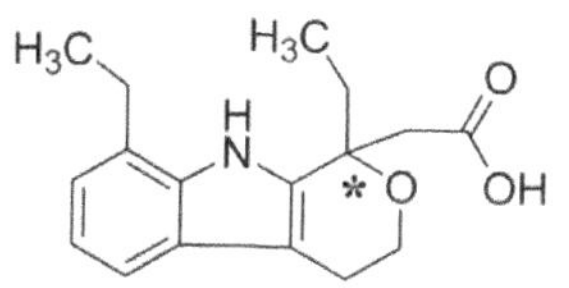

Etodolac

Arylacetic and arylpropionic acids

The latter represent another group of derivatives.

General structure of arylacetic and arylpropionic acids

Ar is an aromatic or heterocyclic residue (thiophene, indole, phenothiazine, etc.).

They have the general indications and contraindications of NSAIDs.

Arylacetic acids are Ar-CH_2-COOH (Figures 3.21 and 3.22).

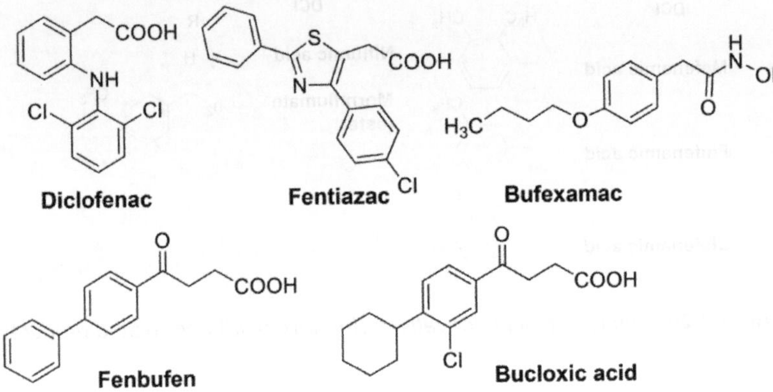

Figure 3.21. *Arylacetic derivatives and precursors*

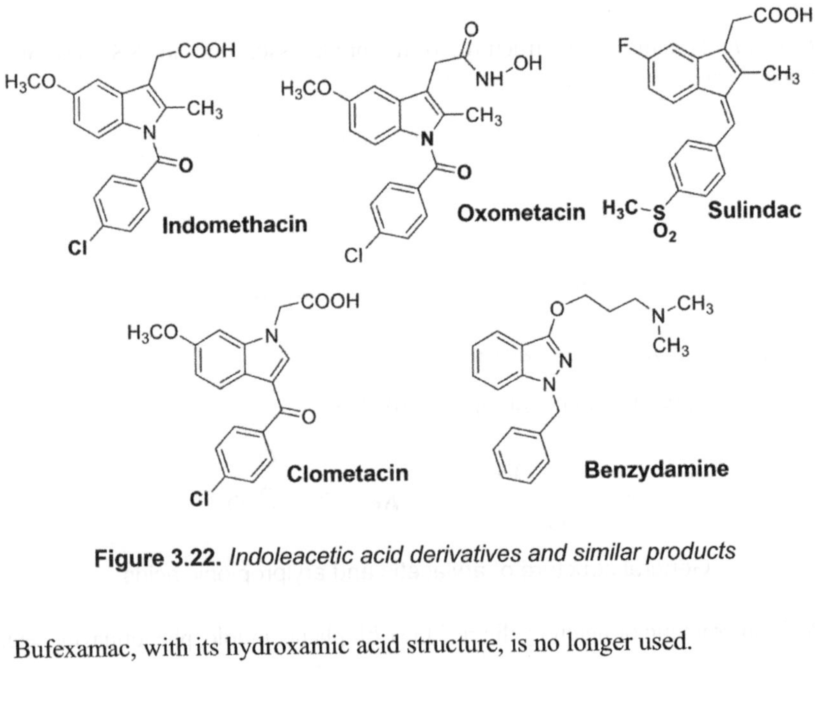

Figure 3.22. *Indoleacetic acid derivatives and similar products*

Bufexamac, with its hydroxamic acid structure, is no longer used.

Diclofenac is a homoanthranilic acid derivative. Bufexamac, with its hydroxamic acid structure, is used to be prescribed exclusively as a topical agent, but is no longer used in France. *Fentiazac* is a thiazole derivative. *Fenbufen* is a precursor transformed in vivo into arylacetic acid. These last two products are rarely used:

– *Indometacin* is an indole derivative.

– *Oxametacin* is derived from a hydroxamic acid.

– *Sulindac* is an Indometacin analogue.

– *Clometacine*. Analgesic activity – withdrawn from the market due to risks of acute and chronic hepatitis.

– *Benzydamine* – vaginal solutions – mouthwash.

Indications: inflammatory rheumatism, spondylitis, polyarthritis, gout.

Contraindications: active peptic ulcer, history of ulcer, rectitis, rectorrhagia.

Drug interactions: associations with other NSAIDs, oral anticoagulants, sulfonamides, Diflunisal (risk of fatal hemorrhage).

– *Nepafenac* has an amide structure, a precursor of an arylacetic acid.

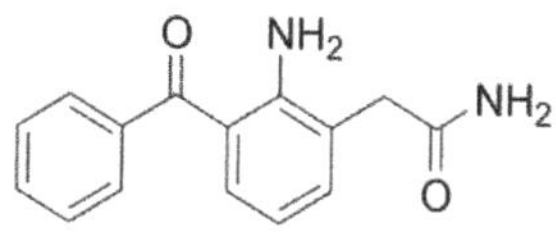

Nepafenac

This NSAID is indicated for the prevention and treatment of pain and inflammation following cataract surgery (removal of the lens of the eye).

2-Arylpropionic acids, also known as "propionics"

The structure of these acids features a chiral center with R and S-enantiomers.

The typical drug is *Ibuprofen*, whose racemate is currently in use. The active derivative is the S (+) enantiomer.

The main representatives are shown in Figure 3.23.

They have the general indications and contraindications of NSAIDs.

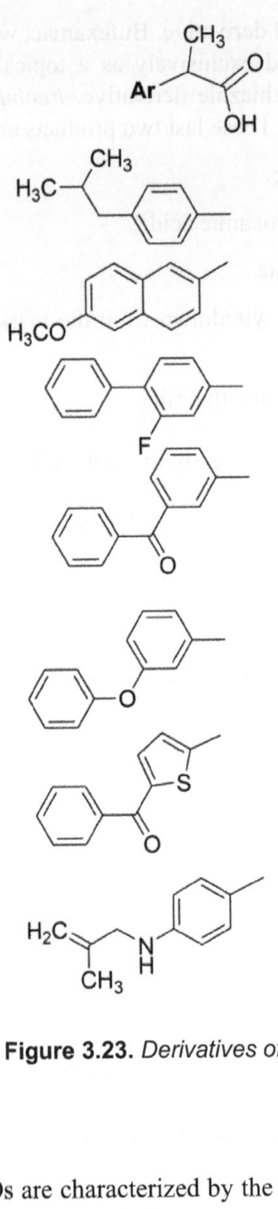

CID

Ibuprofen

Naproxen

Flurbiprofen

Ketoprofen

Fenoprofen

Tiaprofenic acid

Alminoprofen

Figure 3.23. *Derivatives of a 2-arylpropionic acid*

Oxicams

These NSAIDs are characterized by the presence of a carboxamide function and have an acidic character due to the presence of an enolic hydroxyl group.

The main representatives are *Piroxicam*, *Tenoxicam* and *Meloxicam*, which have the same indications and contraindications as other NSAIDs.

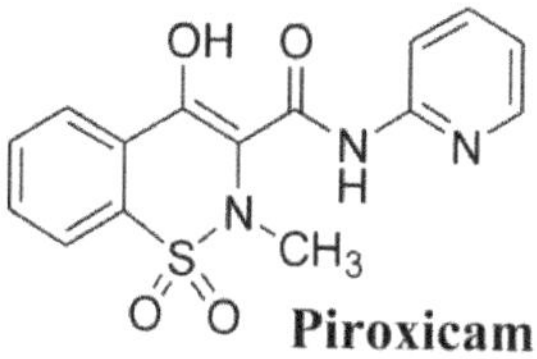

Piroxicam

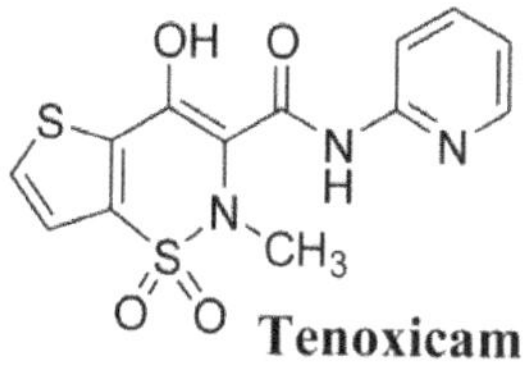

Tenoxicam

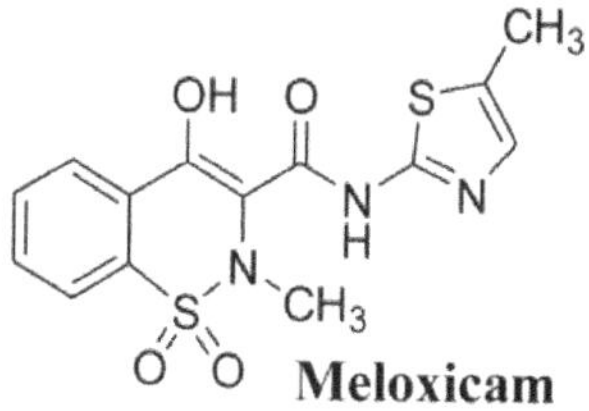

Meloxicam

NOTE.– *Phenylbutazone*, a pyrazolidine derivative with an acidic enol group, is still used in some countries. This compound corresponds to the 4-n-butyl-1,2-diphenylpyrazolidine-3,5-dione with enolic forms. Phenylbutazone and oxicams have the pharmacological properties of other NSAIDs. They are used in the treatment of chronic progressive polyarthritis, ankylosing spondylitis and for their uricosuric properties.

3.1.4.2.2. Selective COX-2 inhibitors (Figure 3.24)

Anti-inflammatories specifically inhibiting COX-2 are derivatives with better gastric tolerance. They do, however, require closer cardiovascular monitoring. Unlike non-selective NSAIDs, they have no anti-platelet effect. They lead to the increased production of thromboxane A2 and leukotriene B4. This imbalance is at the root of the increased risk of thrombosis and cardiovascular events. As a result, *Rofecoxib* (formerly VIOXX®) and *Valdecoxib* have been withdrawn from the market.

The main representatives are shown in Figure 3.24. Their structure includes either a sulfonamide function or a methylsulfone function. The presence of the sulfonamide residue in some derivatives may give them a weak acid character due to tautomerism.

Nimesulide is an inhibitor of COX-2-dependent prostaglandin formation, with weak COX-1 inhibitory activity.

It is reserved solely for the treatment of acute pain and primary dysmenorrhea, as a second-line treatment. It is no longer used in the symptomatic treatment of painful osteoarthritis.

Celecoxib, Nabumetone, Parecoxib and *Etoricoxib* also exhibit selective COX-2 inhibitory activity (Talley et al. 2000), and are used in the treatment of painful and inflammatory osteoarthritis.

Precautions and contraindications are the same as for other NSAIDs.

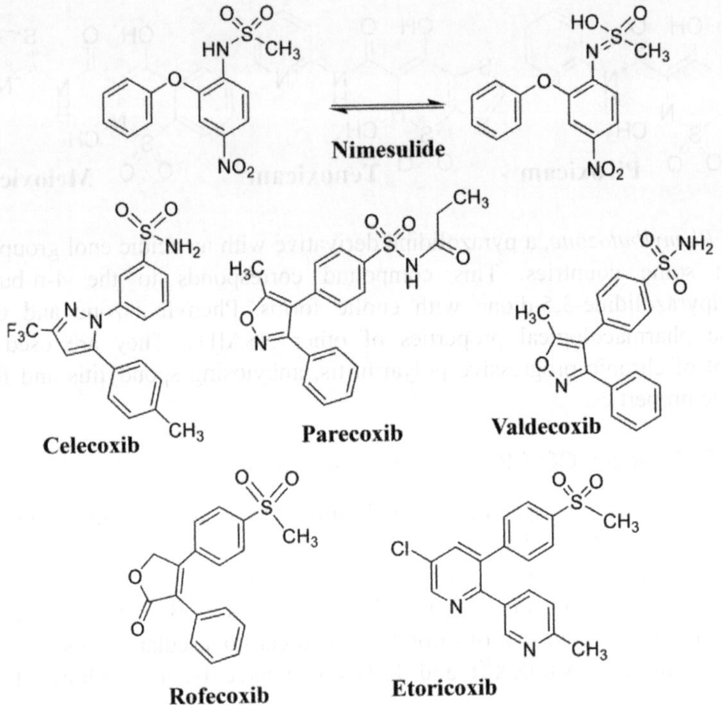

Figure 3.24. *Selective COX-2 inhibitors*

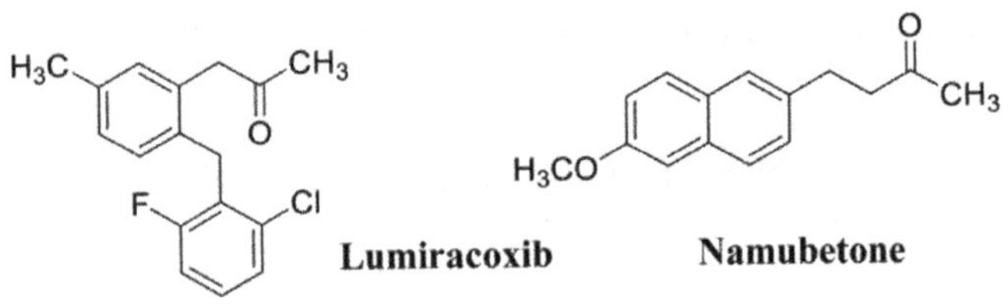

Lumiracoxib is an inhibitor that is similar in structure to NSAIDs of the arylacetic series and anthranilic (fenamic) acids. *Namubetone* is an anti-inflammatory analgesic distinguished from other NSAIDs by its ketone function.

3.1.4.3. *Histidine decarboxylase (EC 4)*

Histidine decarboxylase is an enzyme that decarboxylates histidine. *Tritoqualine* is an inhibitor of histidine decarboxylase, which slows down the formation of endogenous histamine from histidine within tissues.

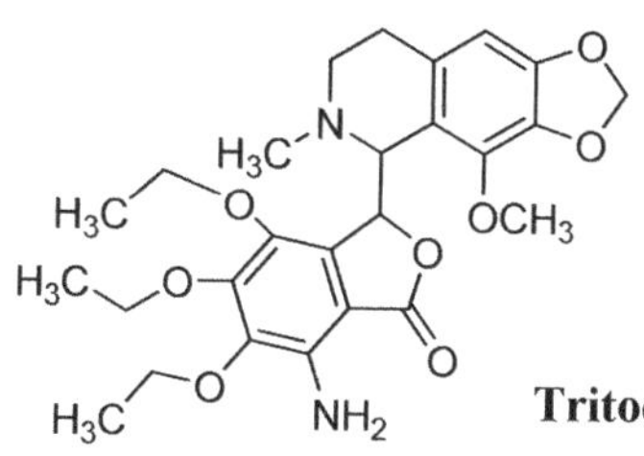

Tritoqualine

Tritoqualine is used in the treatment of certain allergies. It may potentiate hypoglycemic effects. It is contraindicated during pregnancy.

3.1.4.4. *Dihydroorotate dehydrogenase (EC 1)*

This enzyme is involved in the synthesis of pyrimidine bases. It transforms dihydroororate and fumarate into ororate and succinate. A metabolite in *Leflunomide* inhibits the enzyme, providing an anti-proliferative action. Leflunomide is effective in patients suffering from arthritis, psoriatic rheumatism, as well as in certain autoimmune pathologies (rheumatoid arthritis). It is an immunomodulating agent, whose metabolite exerts an anti-proliferative activity against mononuclear cells and T lymphocytes.

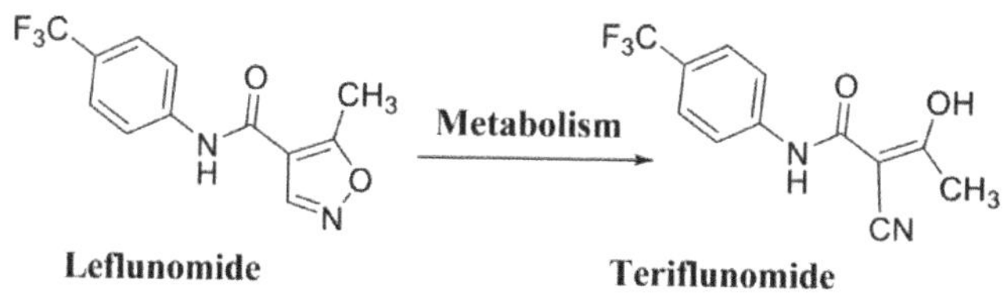

Dihydroorotate dehydrogenase inhibitors are drugs used in the treatment of multiple sclerosis. For example, *Teriflunomide,* a metabolite of Leflunomide, is an immunomodulatory agent that blocks enzyme activity, leading to the inhibition of pyrimidine base biosynthesis. The result is a reduction in pro-inflammatory T lymphocytes. This activity enables the oral treatment of this condition, reducing the recurrence rate, as well as the risk of progression of disability (Doshi and Chataway 2017).

3.1.4.5. *Coagulation factor Xa (EC 3)*

This factor is a serine endoprotease cleaving a peptide sequence after an arginine that is preferentially located at the Ile-(Glu or Asp)-Gly-Arg level. This enzyme is a dimer. The monomers are linked by two disulfide bridges. Active factor X converts prothrombin into thrombin. Thrombin converts fibrinogen into fibrin, which forms the clot. This is part of the cascade of enzymatic reactions involved in hemostasis (stopping bleeding following coagulation). *Apixaban* is a potent, reversible and highly selective inhibitor of the factor Xa active site.

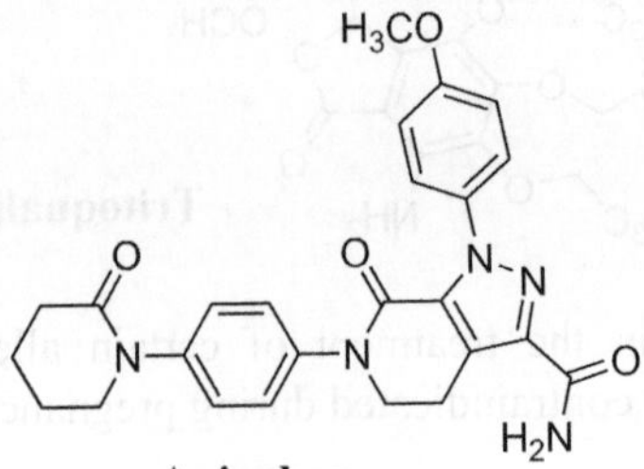

Apixaban

It inhibits this factor, and therefore prothrombinase activity. Apixaban indirectly inhibits platelet aggregation by blocking thrombin function. It is an anti-thrombotic drug used to prevent venous or arterial thrombosis at concentrations that preserve hemostasis. *Rivaroxaban* acts in a similar way to Apixaban.

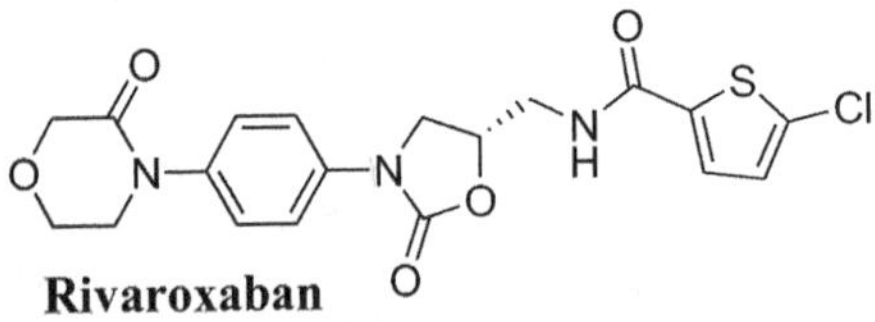

Rivaroxaban

3.1.4.6. *Inosine monophosphate dehydrogenase (EC 1)*

This enzyme is an oxidoreductase that converts inosine monophosphate (IMP) into xanthine monophosphate (XMP) in the presence of NAD^+. This is the initial reaction in the metabolic pathway for GMP synthesis. The catalyst is retro-inhibited by the end product of the pathway, GMP, when it is in excess. Note that the enzyme is capable of catalyzing the reaction in both directions.

Mycophenolic acid is a selective, non-competitive and reversible inhibitor of this enzyme.

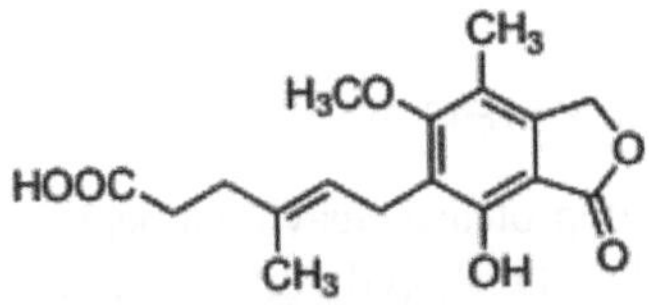

Mycophenolic acid

It inhibits guanine nucleotide synthesis. Given that the development and maturation of B and T lymphocytes are essentially dependent on purine bases, and that other cell types are able to use "secondary" metabolic pathways to produce these molecules, this substance has a cytostatic effect, favorably targeting lymphocytes. This is why it is used as an immunosuppressant in kidney, liver and heart transplants.

3.1.4.7. *Plasma kallikrein (EC 3)*

Kallikrein converts kininogen into kinins. It is an enzyme that can be activated, derived from the cleavage of inactive and circulating prekallikrein by activated factor XII (Hageman). The main kinins are bradykinin and kallidin. They are powerful vasodilators, increasing vascular permeability. *Ecallantide* is a specific, reversible kallikrein inhibitor. It is used in the treatment of angioedema. It is a recombinant protein of 60 amino acids. This drug is marketed in some countries.

Numerous areas of research are currently being explored to inhibit this enzyme in various pathologies (oncology, dermatology, pneumology).

NOTE.– It should be noted that *complement C1 esterase inhibitor* is a serum protein, whose main role is to regulate the classical pathway of complement system (CS) activation. Deficiency in this inhibitor can lead to edema. This molecule inhibits activated C1r and C1s (C1 for Complement Factor 1). The complement system was discovered at the end of the 19th century. At the time, it was described as a key player in the fight against microbes. Since then, knowledge has evolved and it has been established that the CS is part of innate immunity. It enables the host to defend itself against pathogens and eliminate its own cells under conditions of apoptosis. Complement can be activated via three pathways: the classical pathway, the lectin pathway and the alternative pathway. These pathways lead to the activation of factor C3 (Complement factor 3), which enables the formation of a membrane attack complex (MAC). The classical activation pathway is initiated by the binding of C1q to C1r and C1s, two serine proteases. C1q is capable of binding to a wide range of targets, including antibody/antigen immune complexes. This initiating reaction of CS activation leads to the formation of a C3 convertase hydrolyzing C3 into C3a and C3b, followed by a C5 convertase cleaving C5 into C5a and C5b. Both convertases are essential in the fight against pathogens. C3a and C5a are at the origin of an inflammatory response. C3b is involved in opsonization. C5b associates with other complement factors to form lytic pores in the target cell (Daugan et al. 2017). The complement C1 esterase inhibitor is also able to act unspecifically at coagulation level by inhibiting kallikrein, as well as plasmin, trypsin, chymotrypsin and factors XI and XII. This inhibitor is used to treat certain edemas.

3.1.4.8. *Plasmin (EC 3)*

Plasmin is a protease with a wide range of specificities. It attacks fibrin, fibrinogen and coagulation proteins. This enzyme is produced by the liver in the form of plasminogen. It is incorporated into the clot as it forms. Subsequent activation of plasminogen to plasmin enables clot degradation. This reaction is catalyzed by two activators. The first is urokinase, obtained from pro-urokinase. The latter is activated by kallikrein and plasmin, obtained by amplification of fibrinolysis. The second is the tissue plasminogen activator (t-PA), secreted by the vascular wall after trauma.

Tranexamic acid develops an antihemorrhagic action through plasmin inhibition.

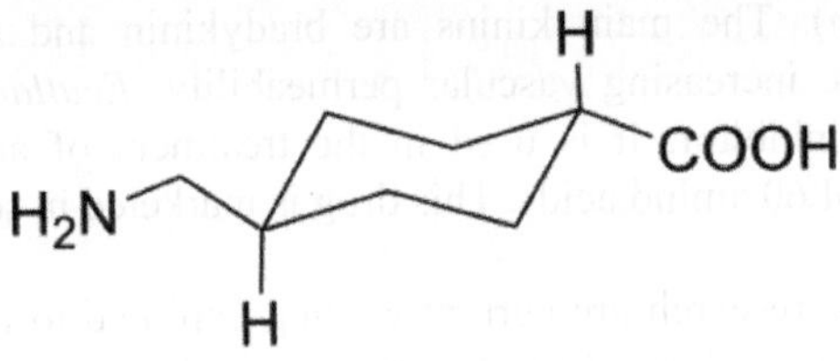

Tranexamic acid

The molecule forms a complex with plasminogen, preventing its transformation into plasmin. The result is the reduced fibrinolytic activity of plasmin. This drug is used to treat fibrinolysis-related bleeding disorders. There are also drugs that act on plasminogen in an opposite way to tranexamic acid. These are synthetic protein molecules corresponding to t-PA, such as *Alteplase,* Tenechteplase and *Reteplase*, which activate plasminogen cleavage into plasmin. These thrombolytics enable the lysis of clots that are implicated in cardiac ischemia leading to, or having caused, infarction.

3.1.4.9. *Thrombin (EC 3)*

This enzyme is a serine protease involved in the coagulation process. This catalyst is formed locally from a prothrombinase enzyme complex. Thrombin is the product of a finely tuned cascade of activations triggered by the meeting of plasma coagulation factors and tissue factor exposed at the site of injury. Thrombin cleaves soluble fibrinogen into monomers, which polymerize to form the insoluble fibrin network.

Argatroban is a synthetic derivative of L-arginine. It is a direct thrombin inhibitor, binding reversibly to thrombin with very high affinity. This drug has an anticoagulant effect.

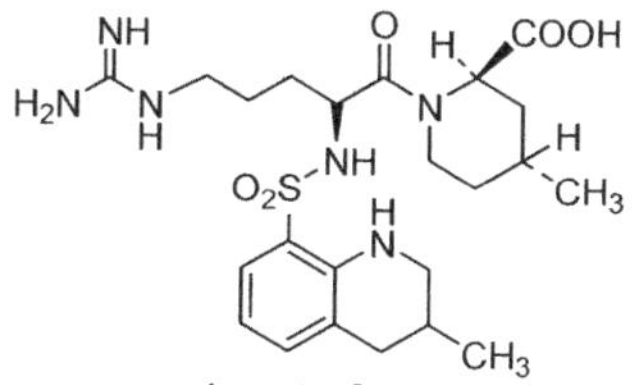

Argatroban

Dabigatran is an orally administered "prodrug" which, upon activation, releases a potent, competitive and reversible thrombin inhibitor.

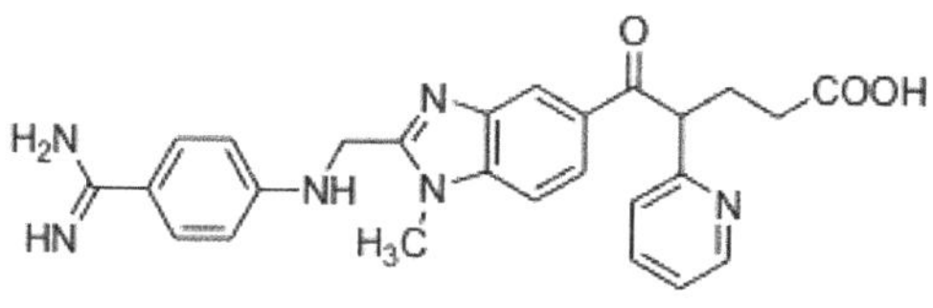

Dabigatran

This medicine is used to treat thrombosis, embolisms and vascular events.

It should also be noted that human thrombin is used in association with human fibrinogen for the management of local bleeding during surgery.

3.1.4.10. *Na^+/K^+ ATPase (EC 3)*

The Na^+/K^+ ATPase pump is a membrane protein complex enabling active cotransport (antiport – using energy from ATP hydrolysis) to move Na^+ ions out of the cell and K^+ ions in.

Cardiac glycosides, such as *Digoxin*, *Digitoxin* and *Ouabain*, are the main inhibitors of Na^+/K^+ ATPase. These molecules act by binding to the extracellular part of the enzyme, inhibiting its function. Digoxin is the only one used therapeutically in heart failure.

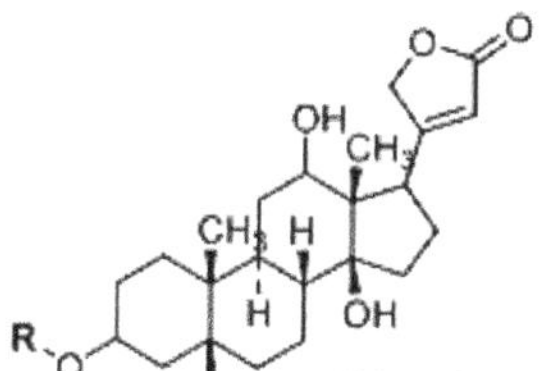

Digoxin

R = Tri-Digitoxose

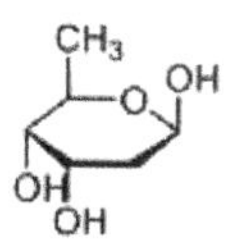

Digitoxose

Other drugs used to treat acute heart failure include *Amrinone*, *Milrinone* and *Enoximone*. These compounds are reserved for hospital use when conventional treatments fail.

3.1.4.11. *Tyrosine 3-hydroxylase (EC 1)*

This enzyme is involved in the formation of catecholamines (dopamine, epinephrine and norepinephrine) from the aromatic amino acid, tyrosine.

Metyrosine, a competitive enzyme inhibitor, acts as an antihypertensive agent.

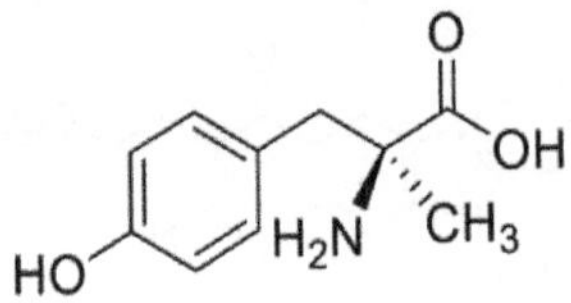

Metyrosine

3.1.4.12. *Vitamin K (epoxy)-reductase (EC 1)*

This enzyme is a complex. It is essential to the vitamin K cycle, a cofactor which, in its reduced form, is essential for the activation of vitamin K-dependent coagulation factors, through carboxylation. This enzyme is inhibited by oral anticoagulants or antivitamin K drugs such as *Warfarin* and *Acenocoumarol*.

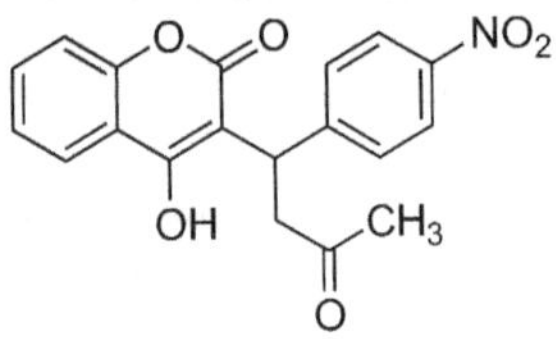

Acenocoumarol

Acenocoumarol and Warfarin are coumarin derivatives that inhibit the enzyme. Acenocoumarol is used to treat embolisms, thrombosis and infarction, often as a replacement for heparin. Warfarin is recommended in the same situations.

3.1.4.13. *Angiotensin-converting enzyme (EC 3)*

Angiotensin-converting enzyme (ACE) is a peptide dipeptidase. It converts angiotensin I into angiotensin II. It has hypertensive properties. Inhibitors of this enzyme (ACE inhibitors) are antihypertensives.

The renin–angiotensin–aldosterone system plays a significant role in regulating blood pressure. Located in the kidney, it ensures water and sodium balance and involves two peptidases: renin, an enzyme of renal origin, and the angiotensin-converting enzyme, secreted by the pulmonary endothelium. They convert angiotensinogen into angiotensin I and the latter into angiotensin II (Figure 3.25).

Angiotensinogen: H_2N-Asp-Arg-Val-Tyr-Ile-His-Pro-Phe-His-Pro-Phe-His-Leu | Val-Ile-His-COOH

Renin ↓

Angiotensin I: H_2N-Asp-Arg-Val-Tyr-Ile-His-Pro-Phe-His-Pro-Phe | His-Leu-COOH

Angiotensin converting enzyme (ACE) ↓

Angiotensin II: H_2N-Asp- | Arg-Val-Tyr-Ile-His-Pro-Phe-His-Pro-Phe-COOH

Peptidase ↓

Angiotensin III: H_2N-Arg-Val-Tyr-Ile-His-Pro-Phe-His-Pro-Phe-COOH

↓

Various inactive metabolites

Figure 3.25. *Action of renin and angiotensin-converting enzyme (ACE)*

Captopril, *Enalapril* and other angiotensin-converting enzyme-inhibiting antihypertensives are used in the treatment of heart failure and arterial hypertension.

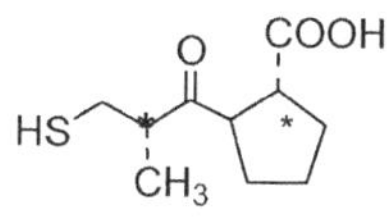

Captopril

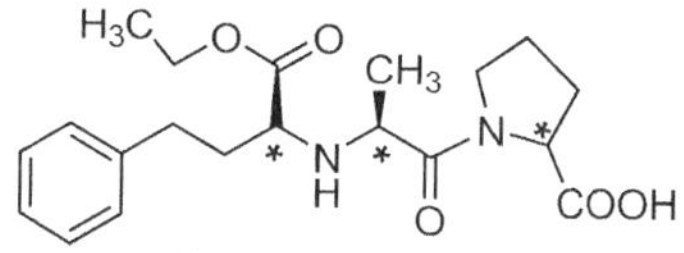

Enalapril

3.1.4.14. *Renin (EC 3)*

Renin functions within the renin–angiotensin–aldosterone system are discussed in section 3.1.4.13. (Figure 3.25). *Aliskiren* is an enzyme inhibitor that is used as an antihypertensive.

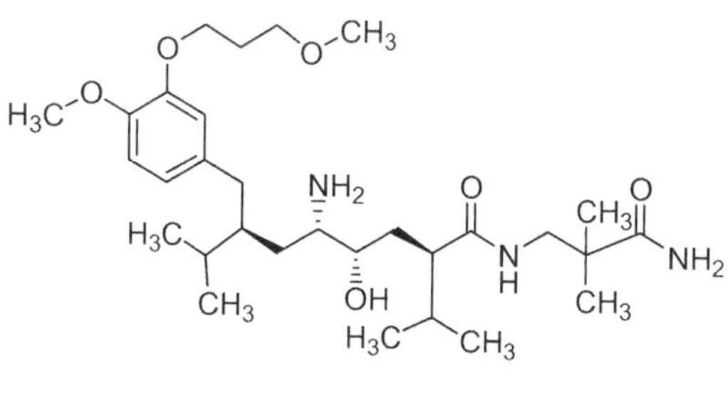

Aliskiren

3.1.4.15. *Nitric oxide synthases (EC 1)*

Nitric oxide synthases, better known as NOS, are enzymes that release the nitric oxide radical NO or nitric oxide (represented by ˙NO).

Three main NOS are known:

– constitutive NOS, comprising neuronal NOS (nNOS or NOS-1) and endothelial NOS (eNOS or NOS-3);

– inducible NOS (iNOS or NOS-2).

NOS are homodimeric enzymes that catalyze the oxidation of L-arginine forming N-hydroxy-arginine in the first step, followed by nitric oxide and citrulline formation in the second (Stuehr et al. 2019).

Neuronal and endothelial NOS produce °NO under the control of calcium (Ca^{2+})-calmodulin. Their monomers contain an N-terminal "oxygenase" domain and a C-terminal "reductase" domain. They catalyze the oxidation of a nitrogen atom in the guanidine group of arginine to nitric oxide (˙NO). Figure 3.26 shows the formation reaction.

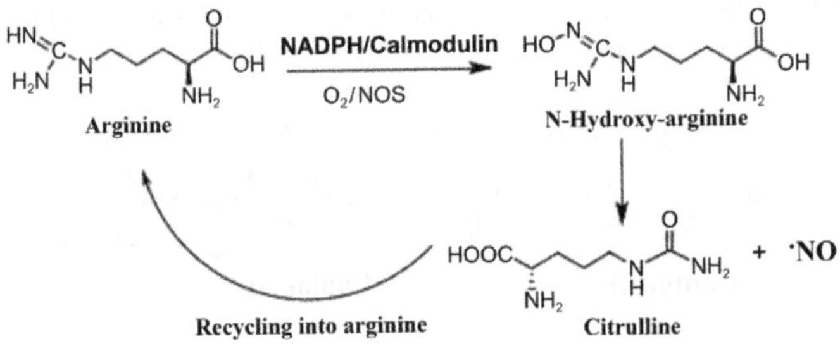

Figure 3.26. *Formation of the nitric oxide radical*

Neuronal NOS1 is implicated in affective disorders (migraines, depression, bipolar disorder and anxiety) and Parkinson's disease. At the nervous level, °NO is a neurotransmitter.

Endothelial NOS3 is the major isoform involved in cardiovascular °NO production. °NO is a vasodilator. Reduced °NO production is a major cause of heart disease, particularly hypertension.

Inducible NOS is mainly involved in inflammatory processes (Pagidipati and Gaziano 2013).

Various pathologies associated with the overproduction or reduced formation of nitric oxide have led to research into the development of derivatives with an action on NOS. Few drugs acting on these enzymes have been marketed to date. However, the search for iNOS inhibitors has been considered in the prevention of hypotension observed during dialysis of patients suffering from renal failure (Nishimura et al. 2000; Sarkar et al. 2004).

Tilarginine corresponding to N-methylarginine, with NOS inhibitory properties, has been proposed for the treatment of breast cancer (Dave et al. 2016).

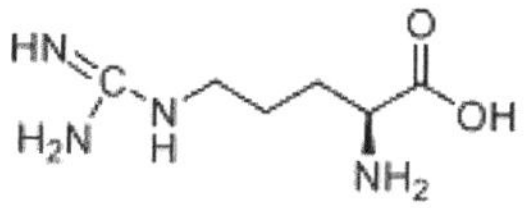

Arginine

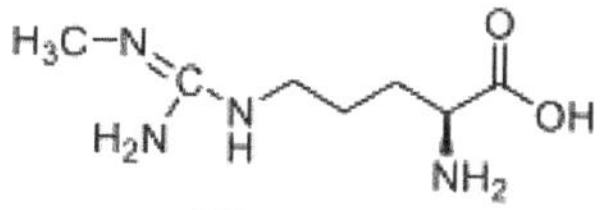

Tilarginine

More °NO-releasing compounds are used therapeutically. These include nitric esters of glycerol (*Trinitrine*) and certain polyols (such as Isosorbide Dinitrate), *Molsidomine* and *Nicorandil* (Tarkin and Kaski 2016), used in the treatment of angina pectoris. They have also been considered for the treatment of intraocular hypertension in glaucoma.

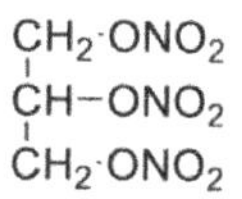

Trinitrine (nitroglycerin)

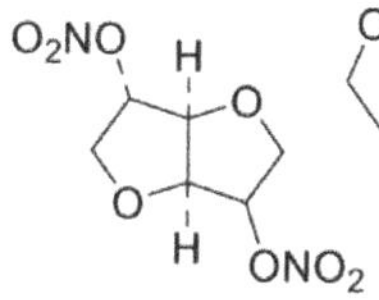

Isosorbide dinitrate

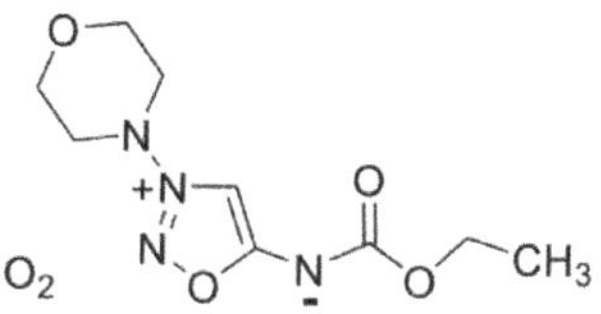

Molsidomine

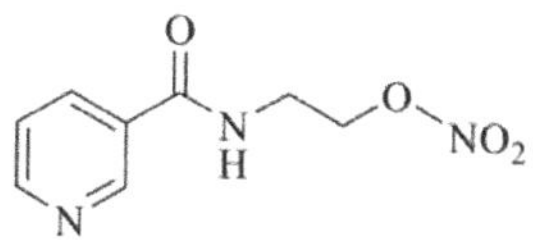

Nicorandil

Numerous applications are currently under development (Sharma et al. 2018).

3.1.5. *Oncology*

3.1.5.1. *Adenosine deaminase (EC 3)*

Adenosine deaminase, an intracellular purine recycling enzyme, is involved in the irreversible hydrolytic deamination of adenosine to inosine, and deoxyadenosine to deoxyinosine (Figure 3.27).

Hairy cell leukemia is a rare chronic B-cell malignancy characterized by the presence of larger-than-normal lymphocytes, with fine chromatin, a nucleolus and hairy cytoplasm. The neutropenia and monocytopenia accompanying this condition are responsible for pulmonary infections, particularly those due to atypical mycobacteria. Treatment of this disease was limited to the long-term administration of interferon alpha (90% patient survival after four years), which stimulates NK (natural killer) lymphocytes and induces differentiation of leukemic cell lines. However, treatment with interferon is characterized by numerous adverse effects (asthenia, myalgia, arthralgia, peripheral neuropathy), hence the interest in *Pentostatin* and *Cladribine*, which are new structural analogues of purine nucleosides (Figure 3.28).

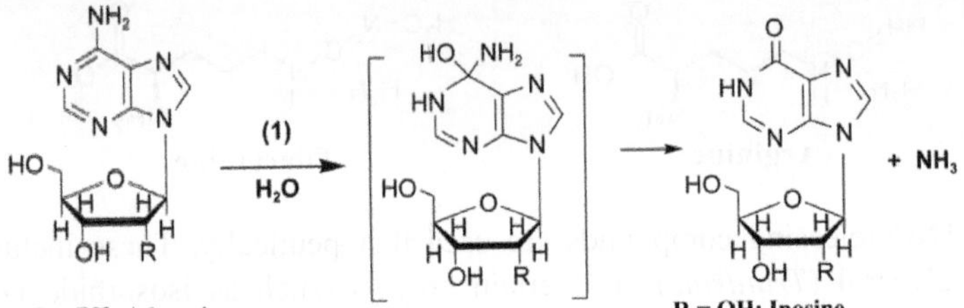

R = OH: Adenosine
R = H: Deoxyadenosine

R = OH: Inosine
R = H: Deoxyinosine

(1): Adenosine deaminase

Figure 3.27. *Mode of action of adenosine deaminase inhibitors*

When phosphorylated by a kinase, Cladribine can disrupt DNA synthesis. This molecule is used to treat rare tricholeukocyte leukemia (B lymphoproliferative syndrome).

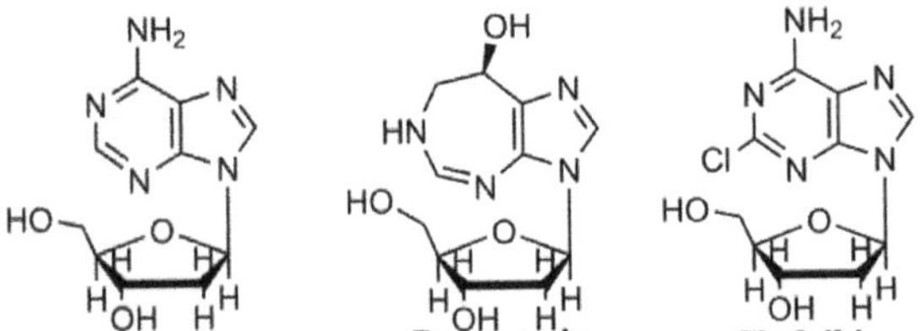

2-Deoxyadenosine | **Pentostatin** | **Cladribine**

Metabolite | Antimetabolites

Figure 3.28. *Structure of deoxyadenosine, Pentostatin and Cladribine*

Pentostatin and Cladribine are adenosine deaminase inhibitors (Figure 3.28). In the absence of this enzyme, unnatural nucleoside analogues accumulate, and then unnatural nucleotide analogues become toxic, leading to leukemia cell death. Cladribine is a structural analogue of deoxyadenosine, with chlorine in position 2 instead of hydrogen. This replacement makes the molecule resistant to enzyme action. It acts as an adenosine competitor.

NOTE.– Adenosine deaminase deficiency causes severe abnormalities in the number and function of various lymphocytes (B (bone), T (thymus) and NK (natural killer)). Work has been carried out on enzyme replacement, and the FDA has approved *Adagen*, a bovine enzyme replacement in cases of severe immunodeficiency.

3.1.5.2. *Dihydrofolate reductase (EC 1)*

The bacterial homologue of this enzyme has already been described. The function of the human protein remains unchanged. Dihydrofolate reductase produces tetrahydrofolate from dihydrofolate and NADPH. In mammals, dihydrofolate reductase is also the enzyme required for the biosynthesis of purine nucleotides and thymidylate from deoxyuridylate and, consequently, DNA (Figure 3.29). The reaction forms 5,6,7,8-tetrahydrofolic acid, followed by the formation of N_5,N_{10}-methylenetetrahydrofolic acid. The latter is involved in the transfer of the methyl group to 2'-deoxyuridylate to form deoxythymidine phosphate under the action of thymidylate synthase TSase (EC 2.1.1.45) (Figure 3.29). Tetrahydrofolate is involved in the transport of monocarbon radicals. Blocking this transfer will prevent the formation of deoxythymidine monophosphate and other nucleotides that are essential for DNA biosynthesis. The result is the stop of cell division (Figure 3.29). Recent work completes this formation mechanism with the involvement of other enzymes (Mishanina et al. 2012).

The first dihydrofolate reductase inhibitor introduced into therapeutics was *Aminopterin*, whose toxicity led to its discontinuation. It was followed by *Methotrexate* (MTX), an antimetabolite that competitively inhibits this enzyme, which is necessary for the biosynthesis of purines and pyrimidines, and thus of DNA. More recent derivatives with this property have been introduced into therapeutics (Figure 3.30). Their action is potentiated when combined with 5-fluorouracil (complementary process) (Figure 3.29).

Methotrexate (Figure 3.30) is used to treat a variety of cancers: choriocarcinomas, breast and ovarian cancers, small-cell lung cancer, upper aerodigestive tract cancers, acute lymphoblastic leukemia (maintenance therapy). At high doses, it is recommended for acute lymphoblastic leukemia in children, and malignant non-Hodgkin's lymphoma.

Side effects include hematological toxicity (thrombocytopenia, leukopenia, even pancytopenia), nausea, vomiting, liver and kidney toxicity, and drug interactions with non-steroidal anti-inflammatory drugs (NSAIDs), especially pyrazole and salicylate derivatives, Cotrimoxazole, Phenytoin, phenobarbital and Ciclosporin.

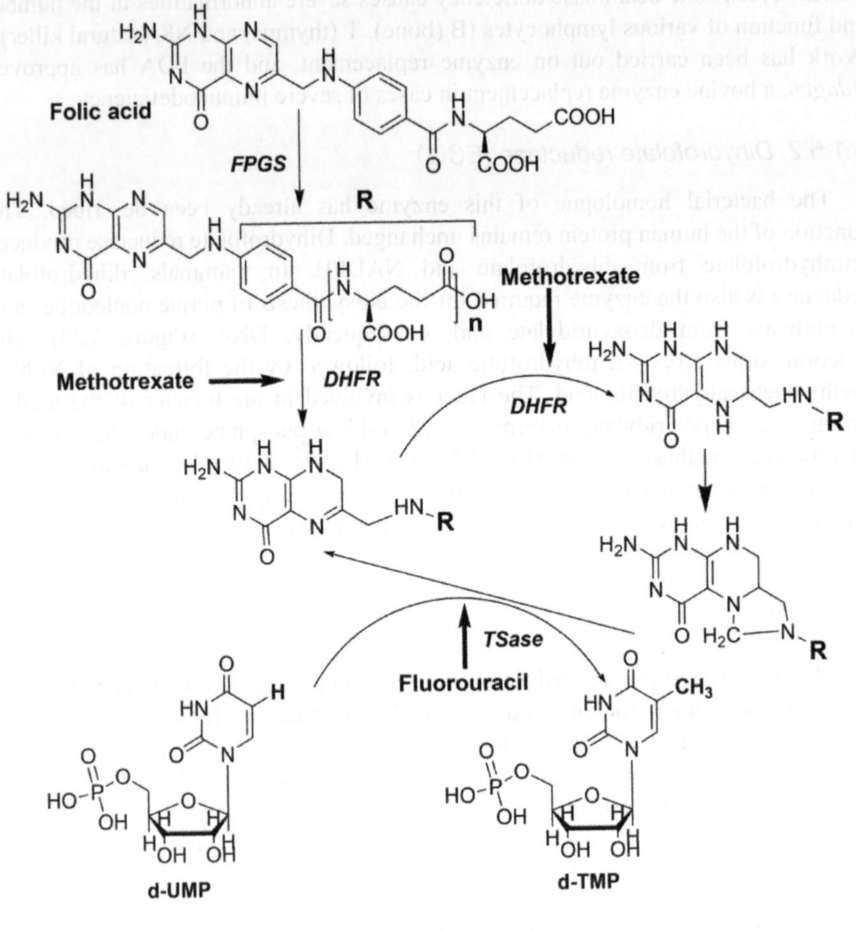

Figure 3.29. *Mode of action of Methotrexate and 5-fluorouracil. FPGS: folylpolyglutamate synthase; DHFR: dihydrofolate reductase; TSase: thymidylate synthase*

Other derivatives (Figure 3.30) are the following:

– *Pemetrexed* inhibits DHFR, thymidylate synthase and glycinamide ribonucleotide formyl transferase. It is mainly used to treat non-small cell lung cancer.

– *Ralitrexed* inhibits thymidylate synthase and other enzymes. It is used to treat lymphomas.

– *Pralatrexate*, similar in structure to Aminopterin but with the loss of nitrogen in 10, inhibits DHFR and folylpolyglutamyl synthetase. It is used to treat refractory T-cell lymphomas that are resistant to Methotrexate.

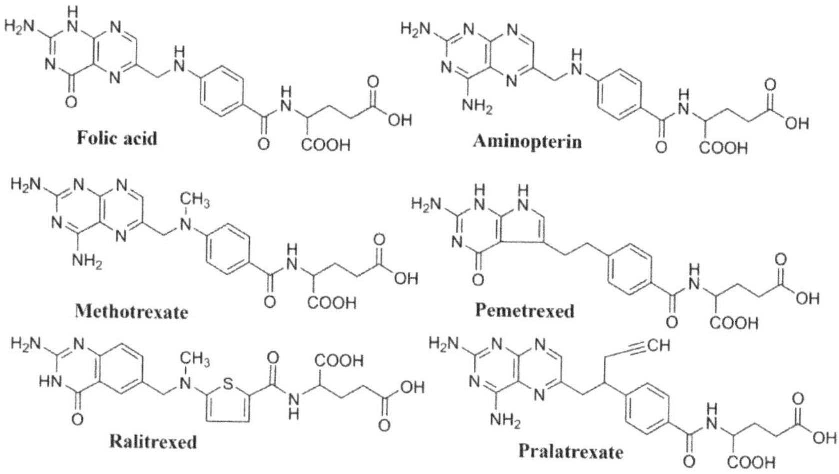

Figure 3.30. *Structure of folic acid and DHFR inhibitors*

3.1.5.3. *Thymidylate synthase (EC 2)*

This enzyme has already been discussed in terms of targeted catalysts in mycoses, sulfonamide antimicrobials and, in section 3.1.5.2, on human dihydrofolate reductase. It is involved in thymidine synthesis and the formation of DNA molecules (Figure 3.31).

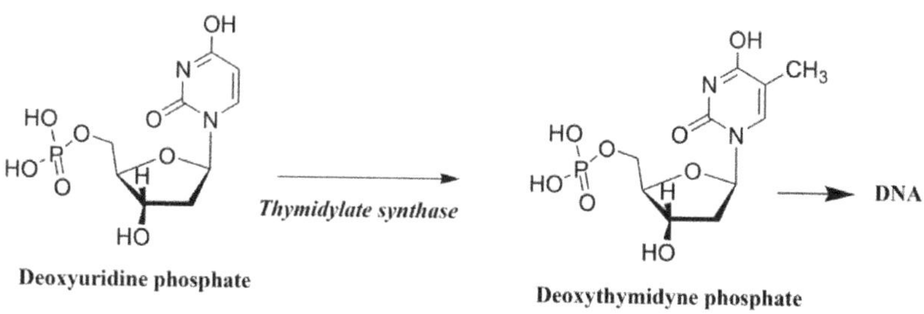

Figure 3.31. *Action of thymidylate synthase*

It catalyzes the transfer of a monocarbon radical onto deoxy-UMP to form deoxy-TMP. *5-Fluorouracil* is an antimetabolite, a competitive inhibitor of thymidylate synthase. It also inhibits uridine phosphorylase.

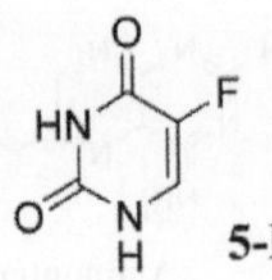

5-Fluorouracil

As already mentioned, Pemetrexed is a multi-targeted anti-folate antineoplastic agent that acts by interrupting folate-dependent metabolic processes that are essential for cell replication. This agent inhibits thymidylate synthetase, dihydrofolate reductase and glycinamide ribonucleotide formyltransferase, which are key folate-dependent enzymes for the de novo biosynthesis of thymidine and purines. The molecule is used in the treatment of lung cancers.

3.1.5.4. *3-Oxo-5-alpha steroid 4-dehydrogenase (EC 1)*

This enzyme, also known as 5-alpha reductase, carries out three main reactions. These are the conversion of testosterone into 5-alpha dihydrotestosterone, and the transformation of progesterone and corticosterone into their 5-alpha oxosteroid derivatives. This enzyme plays a major role in sexual differentiation. The discovery of two isoforms of this enzyme (type 1 and type 2) has paved the way for the development of new therapeutically relevant compounds.

The type 1 isoform is mainly found in the liver, and is involved in the metabolism of androgens and testosterone. The type 2 isoform acts in the prostate and testicles.

The discovery of the absence of benign prostatic hypertrophy in castrated subjects, or those with a genetic deficiency of this enzyme, led to the development of 5-alpha reductase inhibitors.

Dihydrotestosterone (DHT) targets receptors on prostate cells and plays a major role in prostate development and enlargement. Inhibiting this enzyme prevents DHT from stimulating these receptors. *Finasteride* and *Dutasteride* (Figure 3.32) are azasteroids that inhibit NADPH-dependent 5-alpha reductase.

Their action blocks the formation of dihydrotestosterone, whose target is prostate tissue. They induce a regression of the hyperplastic process and, in the long-term, a reduction in the volume of the gland, with the regression of obstruction and a reduced risk of urinary retention. They are used in the treatment of benign prostatic hyperplasia. Unlike anti-androgens and GnRH analogues, they do not interfere with testosterone's sexual functions. Anti-androgens are used to treat hormone-dependent prostate cancer. In addition to steroidal and non-steroidal anti-androgens, natural estrogens, diethylstilbestrol, progesterone and progestins, as well as GnRH analogues, induce an anti-androgenic action.

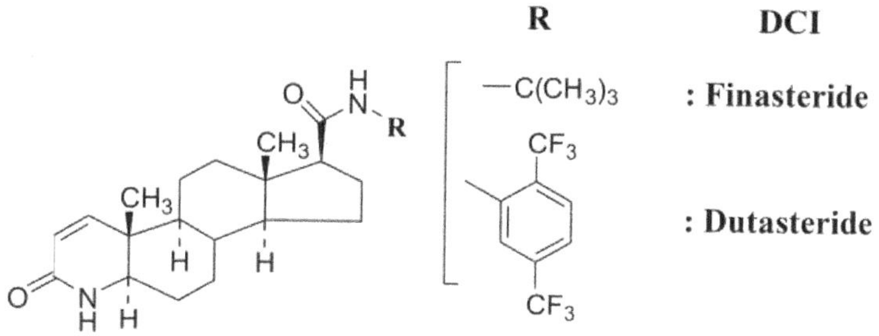

Figure 3.32. *Azasteroid 5-alpha reductase inhibitors*

3.1.5.5. *DNA topoisomerase (EC 5)*

Topoisomerases are DNA-cleaving enzymes that are mainly involved in replication and transcription. Inhibitors of these enzymes form an important group of anti-cancer drugs and antibiotics, whose activity results from the inhibition of bacterial topoisomerases or DNA-gyrases. The DNA molecule is topologically compact and tightly wound, making replication impossible without factors to loosen the structure. Topoisomerases are such factors. They act by transiently cutting one strand of DNA (topoisomerases I) or two strands (topoisomerases II), thus enabling replication. Topoisomerase acts by attaching to a DNA chain via its tyrosine residue at the 3',5' internucleotide phosphodiester bonds (see Figure 2.11 – bacterial topoisomerase I). The result is a break in the DNA with a drop in supercoiling tension. The DNA strand passes through this gap, enabling replication. The next step is religation, with restoration of the DNA strand's previous integrity and separation of the topoisomerase from the DNA. Inhibition of topoisomerases prevents DNA cutting. DNA replication is blocked, interrupting the cell division cycle and leading to cell death. Some drugs act on one or another of these two enzymes:

– *Camptothecin* and its derivatives inhibit topoisomerase I (Figure 3.33).

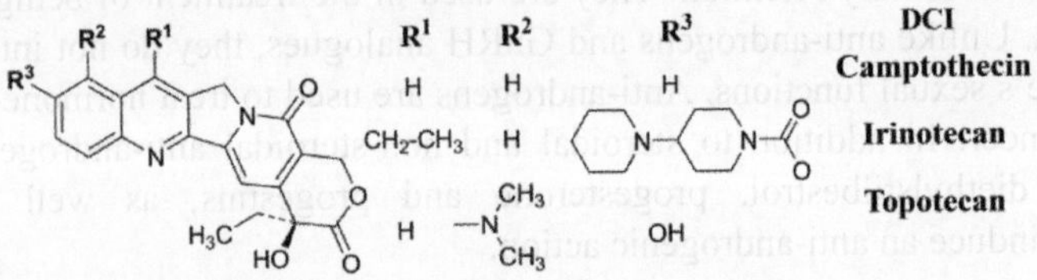

Figure 3.33. *Camptothecin and derivatives*

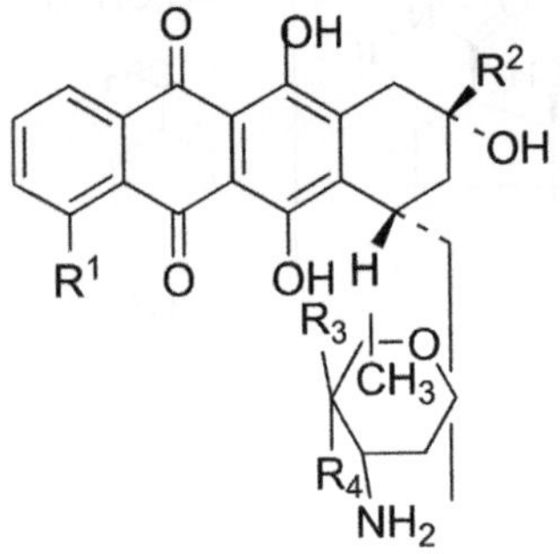

DCI	R¹	R²	R³	R⁴
Daunorubicin Daunomycin, Rubidomycin	OCH_3	$COCH_3$	H	OH
Doxorubicin	OCH_3	$COCH_2OH$	H	OH
Epirubicin	OCH_3	$COCH_2OH$	OH	H
Zorubicin	OCH_3	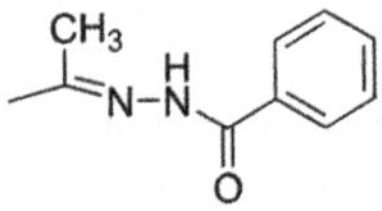	H	OH
Pirarubicin	H	$COCH_2OH$	H	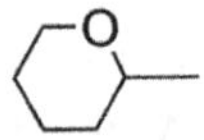
Idarubicin	H	$COCH_3$		OH

Figure 3.34. *Anthracyclines*

– Anthracyclines (*Daunorubicin*, *Doxorubicin*, *Epirubicin*, *Zorubicin*, *Pirarubicin* and *Idarubicin*) (Figure 3.34) and hemisynthetic epipodophyllotoxin heterosides (Etoposide, Teniposide) (Figure 3.35) inhibit topoisomerase II.

These molecules are used in the treatment of many cancers.

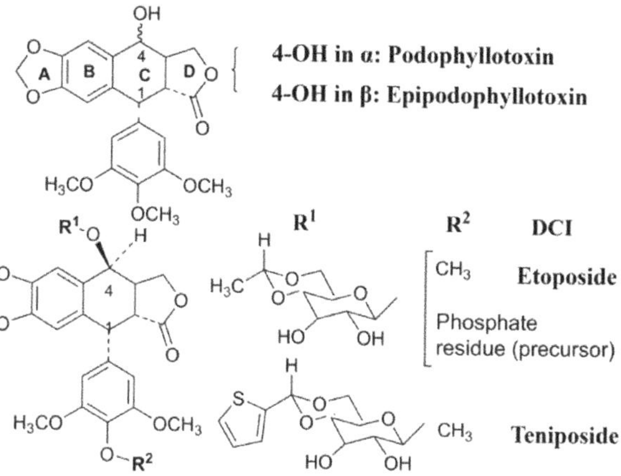

Figure 3.35. *Epipodophyllotoxin derivatives*

3.1.5.6. *Histone deacetylase (various isoenzymes) (EC 3)*

This family of enzymes catalyzes the removal of acetyl groups attached to lysine residues on histones and certain non-histone proteins.

Panobinostat, an antineoplastic agent, is a histone deacetylase (HDAC) inhibitor.

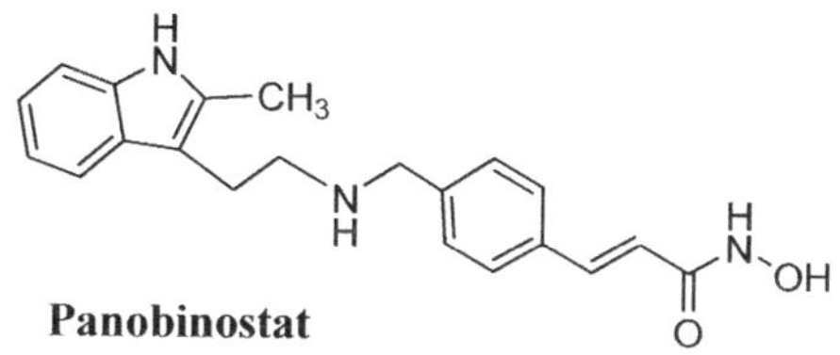

Panobinostat

Inhibition of enzymatic activity leads to increased histone acetylation and activation of transcription of certain genes of interest in certain myelomas. In vitro, Panobinostat leads to the accumulation of histones and other acetylated proteins. This leads to cell cycle arrest and/or apoptosis of certain mutated cells. The substance is more active on tumor cells than on normal cells. This drug is used in the management of relapsed and/or refractory multiple myeloma.

3.1.5.7. *Histone-lysine N-methyltransferase (EC 2)*

The histone-lysine N-methyltransferase, or EZH2 (Enhancer of Zest Homolog 2) enzyme, is encoded by the EZH2 gene. It catalyzes the mono-, di- and trimethylation of histone 3 lysine 27 (H3K27me3) by transferring methyl groups from the cofactor S-adenosyl-L-methionine. It is involved in various biological functions, including transcriptional regulation of hematopoiesis, cell development and differentiation (Yan et al. 2017). Inhibition of this enzyme prevents histone methylation and leads to the inhibition of cell multiplication.

Tazemetostat is a competitive inhibitor of histone S-methyl transferase EZH2, and exerts antitumor activity. It is used for the treatment of epithelioid sarcoma (malignant tumor derived from mesenchymal cells) (Xu et al. 2015).

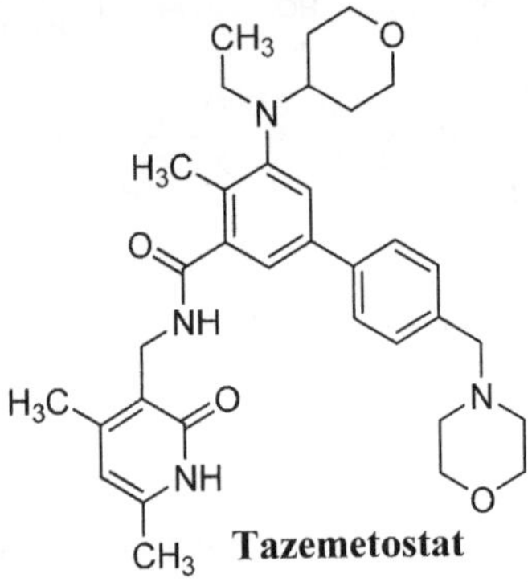

Tazemetostat

3.1.5.8. *Proteasome (EC 3)*

Bortezomib is an inhibitor of this complex. It specifically inhibits the enzyme's chymotrypsin-like activity.

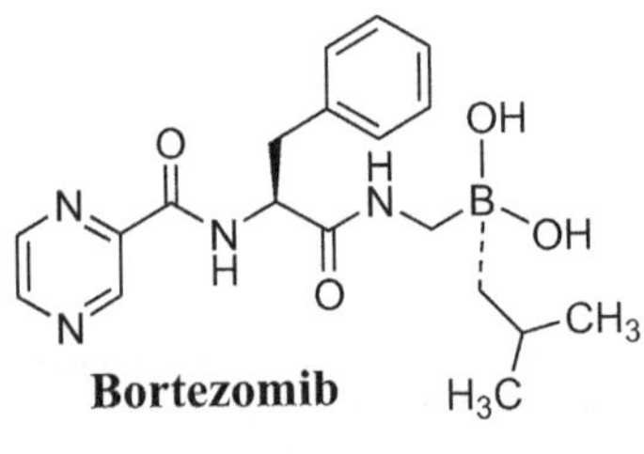

Bortezomib

This inhibition of the proteasome leads to cell cycle arrest and apoptosis. In some myeloma cases, this molecule acts on the tumor cell's ability to interact with the bone marrow microenvironment. Experiments have shown that the substance is particularly active on many cancer cells. These cells are more sensitive to the proapoptotic effects of the molecule than normal cells. In vivo, Bortezomib reduces numerous tumors, including multiple myeloma.

Carfilzomide is a more recent inhibitor derived from a tetrapeptide.

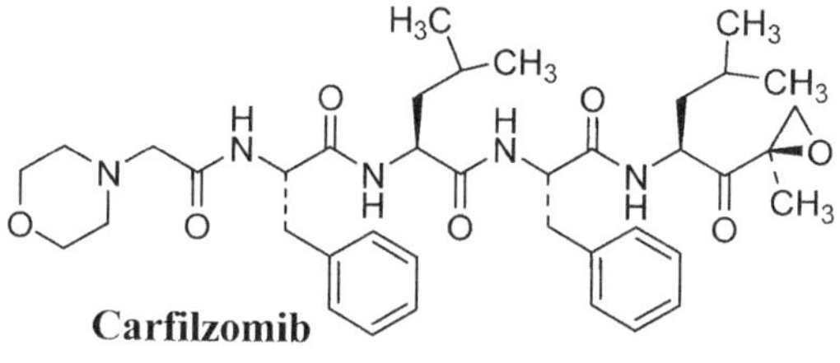

Carfilzomib

It is used as an antineoplastic agent. It binds selectively and irreversibly to the enzyme complex.

Numerous avenues of research are being explored, which involve effectors acting on this enzyme in the context of various pathologies.

3.1.5.9. *Ribonucleoside diphosphate reductase (EC 1)*

In the biosynthesis of purine or pyrimidine nucleotides, this enzyme catalyzes the transition from ribose to deoxyribose, with the loss of a C2' alcohol function. This reaction takes place over several stages involving various molecular and atomic partners: ATP, magnesium, thioredoxin (a hydrogen-transporting protein) and NADPH. Ribonucleoside diphosphate reductase (RNDR) is the enzyme that converts ribonucleoside diphosphates into the deoxyribonucleoside diphosphates required for DNA replication and, consequently, cell division, making it particularly important. Together with a kinase, it is located in the vicinity of DNA polymerase complexes, providing them with the triphosphate deoxynucleoside substrates required for polymerization of the neosynthetized strands.

The RNDR inhibitor *Hydroxycarbamide* prevents the transformation of ribonucleoside diphosphates (ADP, CDP and GDP) into the corresponding deoxyribonucleotides (d-ADP, d-CDP and d-GDP).

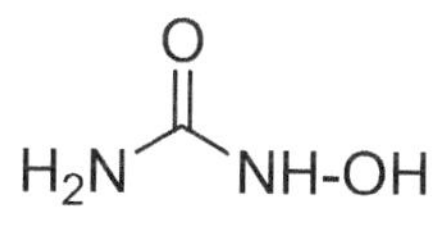

This property gives an antimitotic activity (Figure 3.36), which is thought to result from the chelation of iron required for the enzyme's action during the S phase of the cell cycle.

This derivative is used to treat chronic myeloid leukemia, thrombocythemia, polycythemia and certain cancers. It may give rise to hematological disorders, which should be monitored.

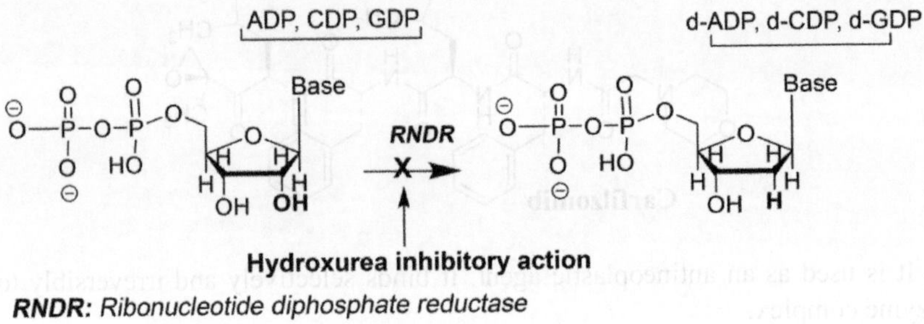

Figure 3.36. *Inhibitory action of hydroxycarbamide*

Clofarabine is a non-metabolizable purine nucleoside analogue.

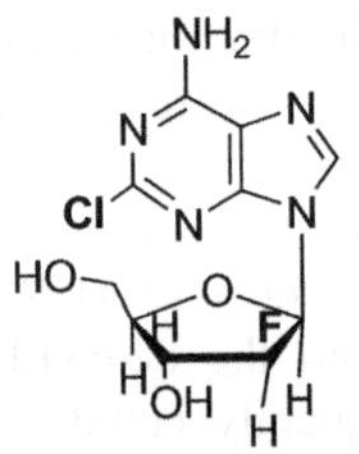

Clofarabine

This molecule has an antitumor effect that is linked to three actions:

– inhibition of ribonucleotide reductase with a reduction in cell concentration of deoxynucleotide triphosphate (dNTP). This is the inhibition discussed here;

– inhibition of DNA polymerase alpha leads to the termination of DNA chain elongation and/or DNA synthesis/repair;

– disruption of mitochondrial membrane integrity with the release of cytochrome c leads to apoptosis, even in lymphocytes that are no longer dividing.

This molecule is used to treat acute lymphoblastic leukemia.

3.1.5.10. *Cytochrome p450 (several isoforms, including aromatase) (EC 1)*

These are involved in the activation of certain drugs, including *Tamoxifen*, used in the treatment of hormone-dependent breast cancer. Cell proliferation is linked to hormonal levels of estrogens in breast cancer and androgens in prostate cancer. Tamoxifen has been the reference anti-estrogen. Its active form appears after metabolism by CYP450, with the formation of 4-hydroxy-tamoxifen and the active metabolite *Endoxifen* (4-hydroxy-N-desmethyl-tamoxifen) (Elkins et al. 2014).

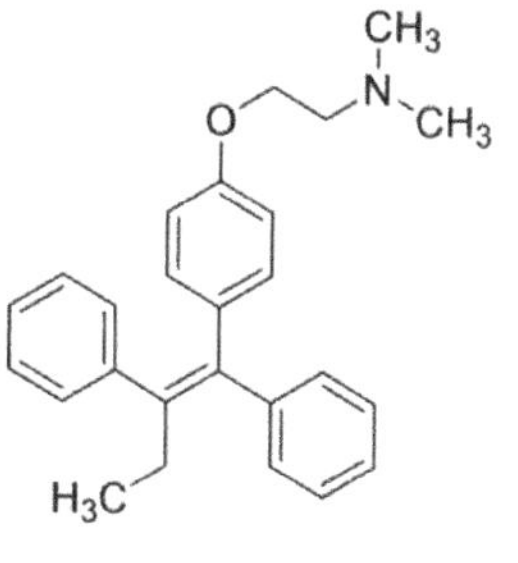

Tamoxifen

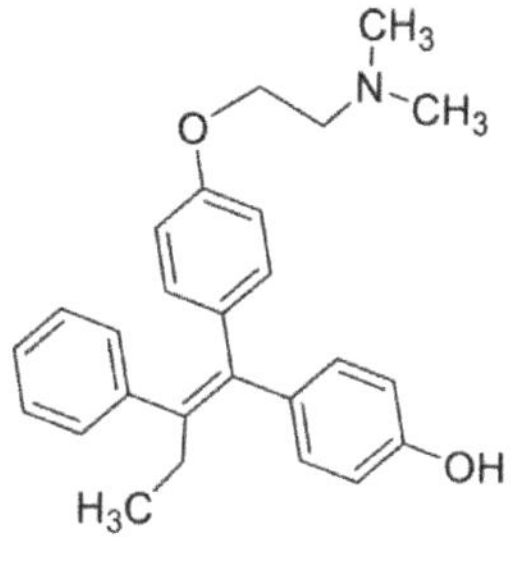

Endoxifen

However, the use of this derivative beyond five years runs the risk of endometrial proliferation and gives rise to various undesirable effects. In these conditions, research has turned to selective estrogen receptor modulators (SERMs), derivatives with modified estrogenicity, which can selectively benefit from the agonist and/or antagonist properties of estrogen. They have therapeutic applications in the treatment and prevention of breast cancer, osteoporosis and post-menopausal cardiovascular disease. Other research has made it possible to avoid Tamoxifen's adverse effects on the uterus, and to treat hormone-dependent breast cancer. Other possibilities have led to the emergence of inhibitors of estrogen biosynthesis, by blocking the activity of aromatase (AI), a cytochrome P450 enzyme (P450 aromatase, CYP 2C19). Action on this enzyme prevents the formation of estrogens from androstenedione and testosterone, without affecting the biosynthesis of other hormones (Figure 3.37).

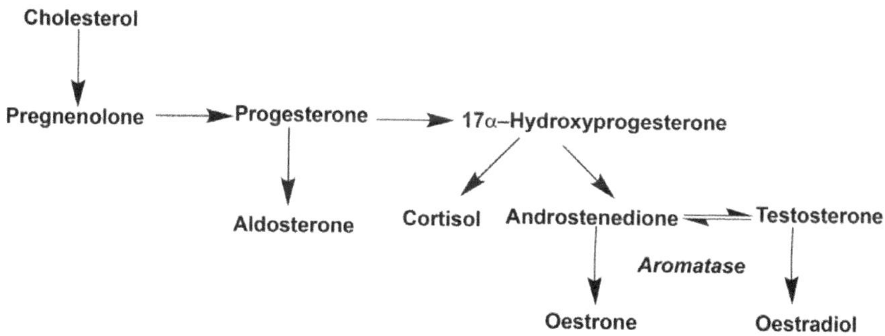

Figure 3.37. *Biosynthesis of steroid hormones and estrogens*

The first to inhibit the conversion of androgens to estrogens was *Aminoglutethimide*. However, its action not only blocks estrogen synthesis, but also that of the mineralotropic hormones and glucocorticoids that are essential to life.

Aminoglutethimide also inhibits other cytochrome P450 enzymes (P450, CYP11 A1), degrading, for example, the cholesterol side chain, leading to other hormones with a steroidal structure.

Aminoglutethimide treatment was abandoned, due to the numerous side effects bringing about more deficient hormones (corticoids). Research has led to more active compounds with selective action on aromatase.

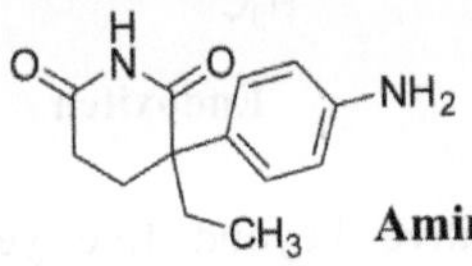

Aminoglutethimide

Current treatments act through selective aromatase inhibition, also blocking adrenal estrogen biosynthesis (Figure 3.37). Since estrogen is mainly secreted by the ovaries, this treatment is reserved for post-menopausal women whose ovaries are no longer functional, in which estrogen is produced mainly in peripheral tissues.

Unlike Aminoglutethimide, these AIs do not block corticosteroid biosynthesis, so there is no need to administer them.

Two groups of AIs are currently used in therapy: compounds with a steroidal structure and those with a non-steroidal structure (Figure 3.38). Their action in the treatment of metastatic breast cancer appears to be more interesting than that of Tamoxifen (Mouridsen et al. 2001; Goss 2003).

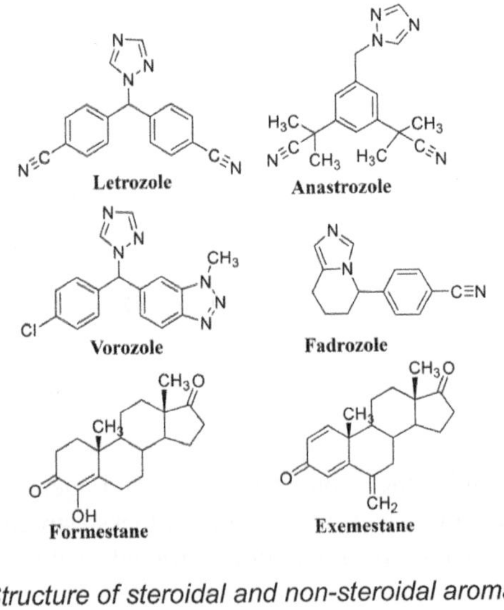

Figure 3.38. *Structure of steroidal and non-steroidal aromatase inhibitors*

They are currently used as first-line treatment for hormone-dependent breast cancer, as well as after Tamoxifen therapy has been discontinued.

Furthermore, the risk of endometrial cancer and thromboembolic disease is lower with AIs, while the risk of osteoporosis and fractures is higher (Baumann and Aebi 2004).

To avoid these risks, AI treatments are combined with biphosphonates or, more recently, monoclonal antibodies (mAbs), inhibiting the ligand of the receptor activator of nuclear factor Kappa B (RANKL), such as Denosumab. It prevents the formation of osteoclasts (responsible for bone demineralization) and helps increase bone mineral density.

They are used after surgery in the preventive treatment of hormone-dependent breast cancer and advanced breast cancer in post-menopausal women.

Cobisistat is a selective inhibitor of the cytochrome P450 3A sub-family. This drug is used to extend the half-life of other drugs that are too rapidly degraded by CYP450s of this sub-family. This compound is given in addition to anti-HIV treatments (Emtricitabine, Elvitegravir, Tenofovir).

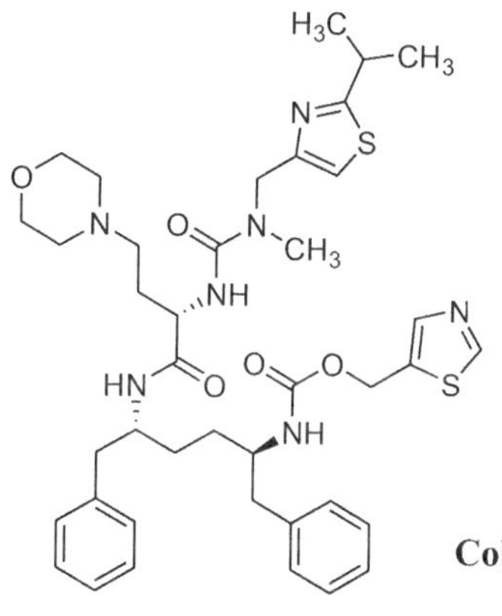

Cobisistat

NOTE.– Some precursors are metabolized and activated by cytochromes p450.

3.1.5.11. *Isocitrate dehydrogenases (EC 1)*

Some leukemias and myelomas are manifested by mutations in isocitrate dehydrogenase type 1 (IDH1) or isocitrate dehydrogenase type 2 (IDH2), leading to the formation of the cancer-inducing metabolite 2-hydroxyglutarate. Inhibition of IDH1 or IDH2 is therefore a treatment for these hematological pathologies, leading to the development of *Ivosidenib* and *Enasidenib*.

– *Ivosidenib* inhibits the mutated form of cytoplasmic isocitrate dehydrogenase 1, and consequently the formation of the cancer-inducing metabolite 2-hydroxyglutarate (2HG). This action leads to the induction of cell differentiation and inhibition of proliferation of tumor cells overexpressing IDH1 (DiNardo et al. 2018).

– *Enasidenib* is an inhibitor of mutated isocitrate dehydrogenase 2, leading to hypermethylation of DNA and histones, with consequent blockage of hematopoietic differentiation. It is indicated in patients suffering from AML (acute myeloid leukemia) and exerts antineoplastic activity (Kim 2017).

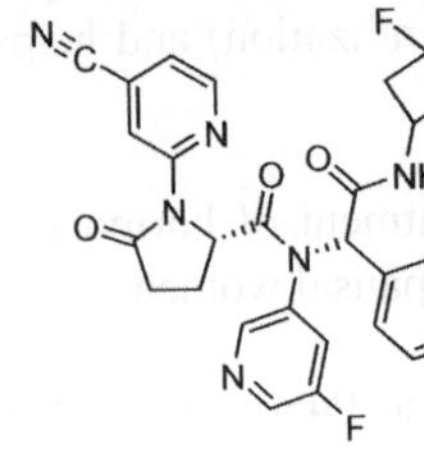

Ivosidenib

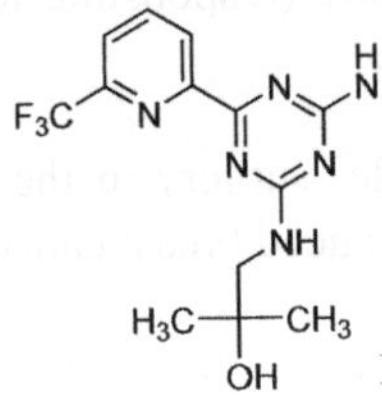

Enasidenib

3.1.5.12. *DNA-dependent DNA polymerases (including telomerases) (EC 2)*

Clofarabine has already been discussed for its inhibition of ribonucleotide reductase. It has also been shown to inhibit DNA polymerase alpha. Among DNA polymerases, the telomerases involved in the replication of chromosome extremities (telomeres) are dysfunctional in a significant number of cancers. Numerous research projects and clinical trials are being conducted on inhibitors acting on these enzymes.

3.1.5.13. *Poly(ADP-ribose) polymerases (PARP) (EC 2)*

Inhibitors of poly(ADP-ribose) polymerases (PARP), 1 (PARP1), 2 (PARP2) and 3 (PARP3) block DNA repair, leading to the accumulation of DNA fragments, genomic instability, cell cycle arrest and apoptosis.

Rucaparib is an inhibitor of these enzymes (Ledermann et al. 2020).

It is a tricyclic derivative with an indole nucleus, and acts as a chemosensitizer/radiosensitizer and antineoplastic.

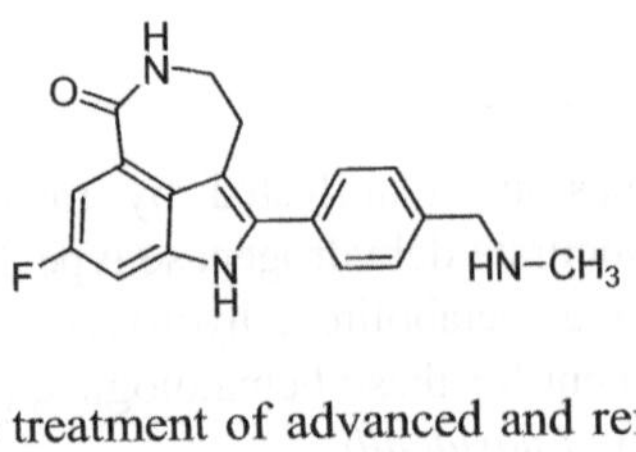

Rucaparib

It is used in the treatment of advanced and refractory ovarian carcinoma. Other indications are also under development.

3.1.6. *Phosphatome and Kinome*

The last two enzyme families yet to be addressed are by far the most extensively studied over the last 10 years. These are the phosphatases and, above all, the kinases (Forrest et al. 2006). So many studies have been carried out that two terms have now been coined: Phosphatome and Kinome. It should be noted that phosphatases are less *druggable* than kinases. They are also less numerous than kinases in a given organism. Research work on them is therefore more time-consuming.

Phosphorylation and dephosphorylation are the main reactions governing signaling pathways (cascades) in the body. While these enzymes are implicated in numerous pathologies, research teams are currently focusing their efforts on oncology.

3.1.6.1. *Phosphatases (EC 3)*

In the body, phosphatases dephosphorylate a wide range of substrates. These dephosphorylations can be either activating or inhibiting. They are involved in numerous cell signaling pathways. There are many dysfunctions and they are at the root of a variety of pathologies. Currently, there are few drugs targeting phosphatases. The broad spectrum of action and, very often, the toxicity of these compounds call for a better understanding of these hydrolases and their roles, in order for them to be targeted efficiently.

Ciclosporin is a drug with immunosuppressive and anti-inflammatory properties. By inhibiting a phosphatase, calcineurin, in dry eyes, this molecule ultimately blocks the release of pro-inflammatory cytokines.

Etidronic acid is the first biphosphonate. It replaces pyrophosphates in the formation of bone tissue. It has an anti-osteoclastic effect and slows bone remodeling (inhibiting bone resorption). It reduces hydroxyprolinuria and the levels of certain phosphatases. This molecule is used to treat osteoporosis, malignant hypercalcemia and Paget's disease, characterized by abnormal and excessive bone remodeling.

Cantharidin is a molecule with inhibitory action on protein phosphatases 1 and 2A, and is used in dermatology.

Protein tyrosine phosphatase 1-B (PTP1-B) is targeted in diabetes and obesity. Molecules directed against this enzyme have reached the clinical phases (Fiedler 2018).

Numerous drug candidates acting on this family of enzymes are in clinical trials or being researched for the treatment of various cancers. Pancreatic acid phosphatase, protein tyrosine phosphatases (PTPs), protein phosphatase 2-A (PP2-A), cdc25 phosphatases, to name but a few, are targets of interest in this field (Korntnera et al. 2018; Kabakci et al. 2019).

3.1.6.2. *Kinases (EC 2)*

In the body, kinases perform the reverse reaction of phosphatases. The term *kinase* is formed from the radical *kin*, derived from the Greek verb *kinein* ("to move") and the suffix *-ase*, which is specific to enzymes. They catalyze the phosphorylation of a wide range of substrates. Their number is much greater than that of phosphatases. As with phosphatases, the catalyzed reaction can be either activating or inhibiting. They are also involved in numerous cell signaling pathways. They have many dysfunctions and are the cause of a wide variety of pathologies. Numerous drugs are currently on the market; it is therefore difficult to give an exhaustive overview. Kinase inhibitors are characterized by their structural diversity. Many of the molecules used in therapeutics are described in sections 3.1.6.2.1 and 3.1.6.2.2.

3.1.6.2.1. Targeted kinases for anti-cancer treatments

A-Receptor tyrosine kinases (RTKs)

In many cancers, mutation or abnormal expression of protein kinases correlates with tumor genesis, metastasis and resistance to chemotherapy. In these conditions, alongside traditional anti-cancer treatments targeting DNA or cell division mechanisms, numerous RTK inhibitors have been developed to treat these pathologies (Robinson et al. 2000). By recognizing and binding to their membrane receptors with tyrosine kinase activity, these drugs take the place of growth factors, preventing the delivery of their proliferation signal. The RTKs mainly targeted are: vascular endothelial growth factor receptor (VEGFR), epidermal growth factor receptor (EGFR), platelet-derived growth factor receptor (PDGFR), hepatocyte growth factor receptor (HGFR), basic-fibroblast growth factor receptor (b-FGFR) and others.

Numerous RTK inhibitor drugs have been brought to market, with a view to treating many cancers (Quan et al. 2017). Various compounds used in therapeutics will be presented, but bear in mind that many molecules are regularly brought to market or under development, given their high therapeutic potential. In a large number of cases, the target that is most representative of a drug's action will be taken into account. The targets reported for these drugs are indicated within the framework of current knowledge, given the existence of multi-targeted compounds.

EGFR

Gefitinib acts by inhibiting the activity of the EGFR receptor. This receptor controls cell survival, cell cycle progression, angiogenesis, migration and invasion in the tumor process. This molecule is mainly used in the treatment of breast cancer and non-small cell lung cancer (NSCLC) (Wakeling et al. 2002).

Erlotinib, an EGFR protein tyrosine kinase inhibitor, is indicated for NSCLC and pancreatic cancer (Abdelgalil et al. 2020).

Some monoclonal antibodies are also used to inhibit EGFR:

– *Cetuximab* is a chimeric monoclonal antibody oriented against the EGFR receptor (Agarwal et al. 2019; Gurdal et al. 2019). It is prescribed for certain colorectal cancers, as well as head and neck cancers.

– *Trastuzumab* is a recombinant humanized monoclonal antibody oriented against human epidermal growth factor receptor 2 (EGFR-2) (Sarosiek and Morawski 2018). It is recommended for the treatment of certain EGFR-2 positive gastric and breast cancers.

EGFR and HER-2

Lapatinib acts on both receptors (Nolting and Schneider-Merck 2014) (Figure 3.39). Its indications are hormone receptor-positive metastatic breast cancer.

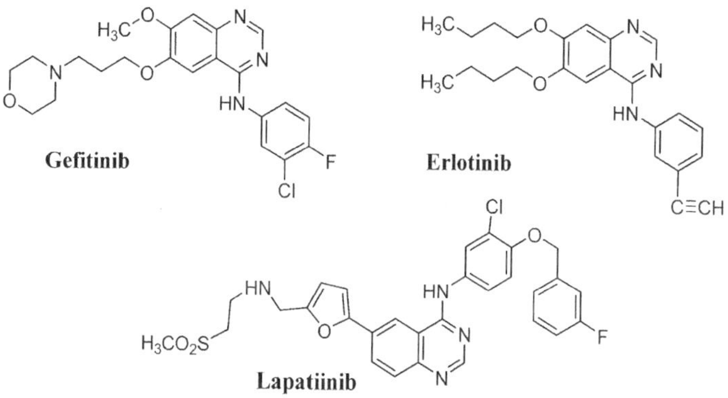

Figure 3.39. *EGFR family receptor tyrosine kinase inhibitors*

ALK

Ceritinib is a potent and highly selective inhibitor of the ALK (anaplastic lymphoma kinase) receptor tyrosine kinase (Massarelli and Papadimitrakopoulou 2015). It inhibits autophosphorylation of the receptor and phosphorylation of proteins downstream of this signaling pathway. It prevents the proliferation of ALK-dependent cancer cells. This drug is used in the treatment of certain bronchial cancers.

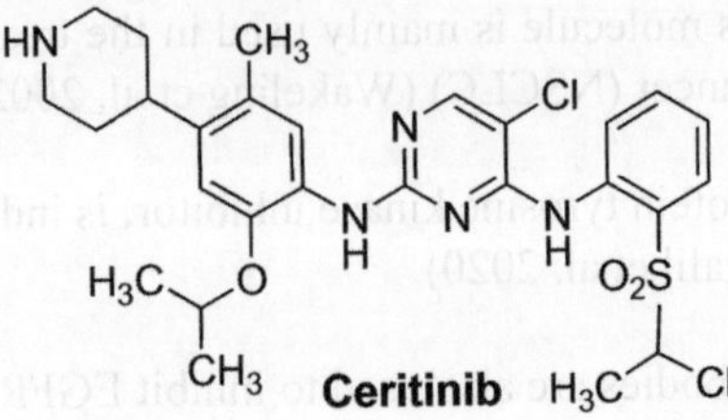

Ceritinib

Trk

Larotrectinib has a substituted pyrazolo-pyrimidine structure.

Its action leads to the inhibition of Trk (a tropomyosin receptor kinase), a receptor tyrosine kinase activated by neurotrophils. It is mutated in various types of cancer. It plays an important role in tumor cell growth and survival. By inhibiting Trk, Larotrectinib binds to the kinase and prevents neurotrophin–Trk interaction and Trk activation. This results in cell apoptosis and inhibition of cell growth in tumors overexpressing Trk (Khotskaya et al. 2017).

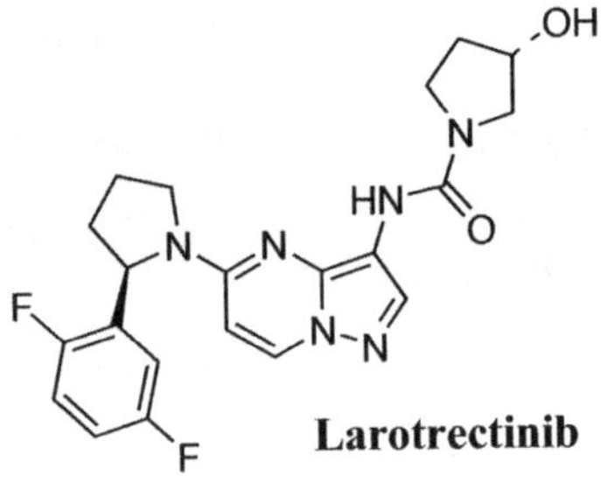

Larotrectinib

VEGFR

Axitinib has a substituted indazole structure (Bellesoeur et al. 2017). It is a potent and selective tyrosine kinase inhibitor of vascular endothelial growth factor receptors (VEGFR-1, VEGFR-2 and VEGFR-3). These receptors are involved in angiogenesis and the neoplastic process. This drug is used to treat kidney cancer.

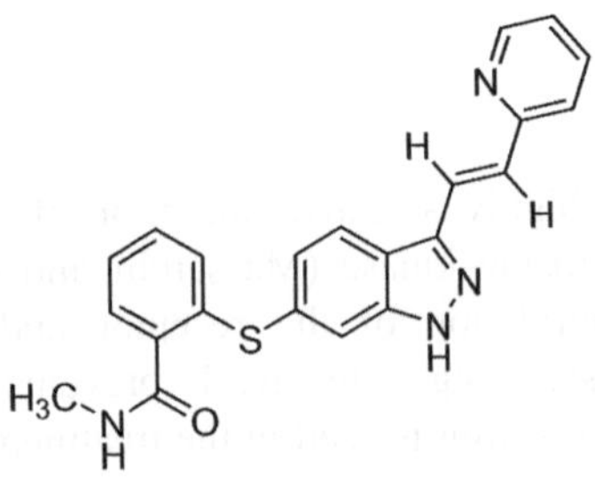

Axitinib

VEGFR and PDGFR

Sorafenib, whose structure corresponds to disubstituted urea, acts by primarily inhibiting the transduction of signals produced by VEGFR or PDGFR (Kim et al. 2017). It is indicated for its actions against renal and hepatic cell cancers.

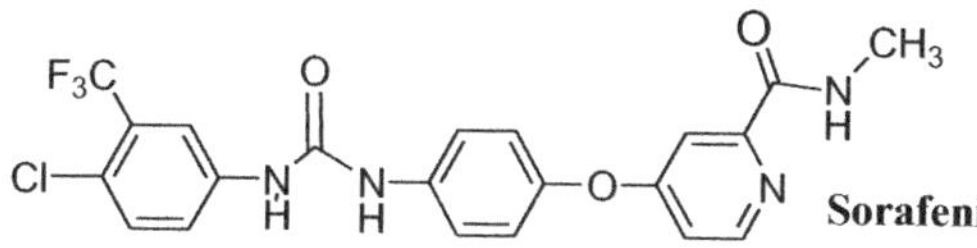

Sorafenib

VEGFR, PDGFR and EGFR

Vandetanib has RTK inhibitory activity, acting against the "cell signaling" processes involved in various cancers (Grande et al. 2013). It targets various receptors: VEGFR, EGFR, PDGFR. Its indications are the treatment of aggressive and symptomatic medullary thyroid cancer in patients with locally advanced non-operable or metastatic disease.

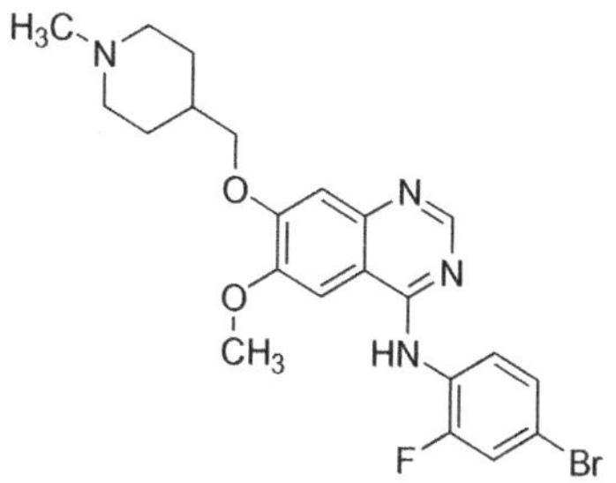

Vandetanib

VEGFR, PDGFR and CSF

Sunitinib inhibits various protein tyrosine kinase receptors, and in particular VEGFR, PDGFR and the stem cell factor receptor CSF (Rizzo and Porta 2017). This derivative is used for its action against gastrointestinal tumors and kidney cancers.

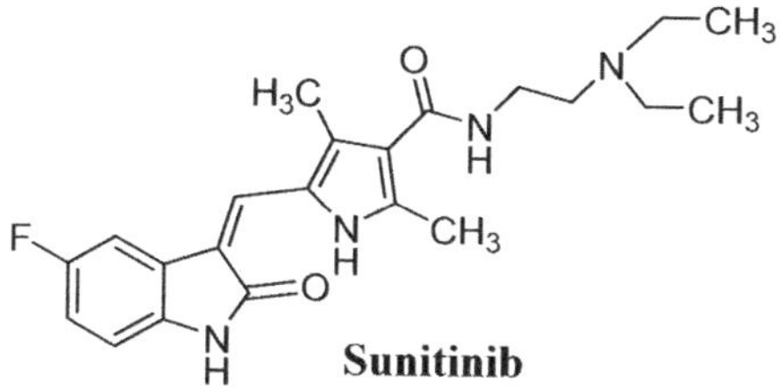

Sunitinib

c-Met, VEGFR2 and RET

Cabozantinib has inhibitory activity against the c-Met (Mesenchymal Epithelial Transition factor receptor), VEGFR2 and RET (REarranged during Transfection receptor) (Kurzroc et al. 2011). It is used in the treatment of medullary thyroid carcinoma.

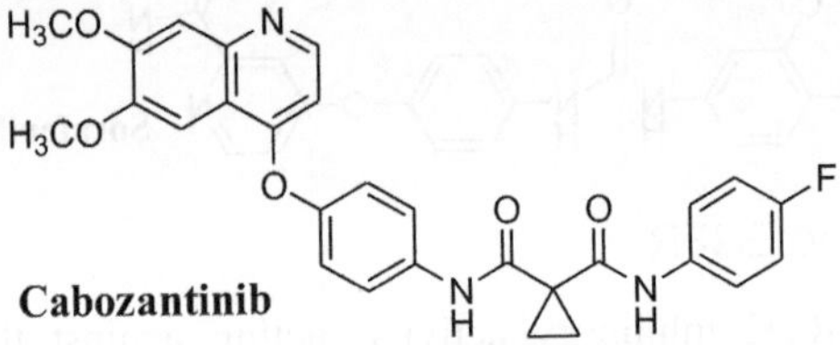

Cabozantinib

ALK and c-Met

Crizotinib is a competitive inhibitor of the receptor tyrosine kinase expressed by the ALK (anaplastic lymphoma kinase) gene and of c-Met tyrosine kinase (Giroux-Leprieur et al. 2016). This molecule inhibits ALK and its mutated variants. It is used in the treatment of non-small cell lung cancer (NSCLC). The risk of severe heart failure (congestive heart failure, low ejection fraction, left ventricular failure, pulmonary edema) should be highlighted.

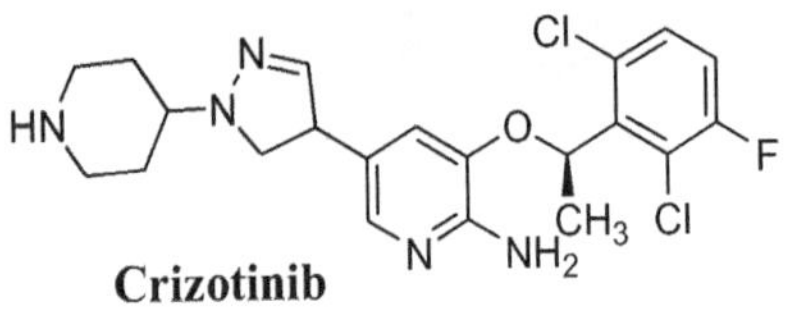

Crizotinib

FLT-3, AXL and ALK

Gilteritinib is a substituted pyrazinecarboxamide, whose formula is indicated.

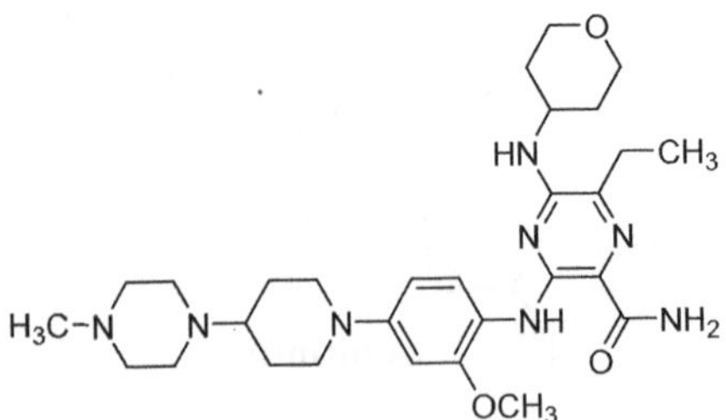

Gilteritinib

The FLT3, FMS-Like Tyrosine kinase 3, gene encodes a tyrosine kinase involved in the proliferation, differentiation and survival of hematopoietic stem cells. Constitutive activation of the FLT3 tyrosine kinase leads to uncontrolled cell proliferation and high white blood cell counts in acute myeloid leukemia (AML). Gilteritinib binds and inhibits wild-type and mutated forms of the FLT3, AXL (receptor tyrosine kinase UFO) and ALK receptor tyrosine kinases. This action inhibits the proliferation of tumor cells overexpressing these receptors or presenting mutated receptors in AML and various types of cancer (Dhillon 2019; McMahon and Perl 2019).

B-Intracellular kinases

Bcr-Abl kinase

Traditional treatment of chronic myeloid leukemia (CML) has involved *Busulfan* (alkylating agent), *Hydroxycarbamide* (hydroxyurea), interferons, allograft and autograft. A chromosomal abnormality, characterized by the absence of a DNA fragment on chromosome 22 (the Philadelphia or Ph chromosome), has been established in CML patients, and has led to changes in the treatment of this disease. The Ph chromosome results from a translocation between two chromosomes, chromosome 9 and chromosome 22. It results from a rearrangement between the Abl (Abelson proto-oncogene) gene on chromosome 9 and the Bcr gene on chromosome 22, leading to two fusion genes:

- Bcr-Abl on chromosome 22;
- Abl-Bcr on chromosome 9.

The Bcr-Abl gene is transcribed into an mRNA that leads to the development of a protein with high tyrosine kinase activity, resulting in cell proliferation and inhibition of apoptosis. Inhibition of the Bcr-Abl protein tyrosine kinase is thus a target for the treatment of CML.

Dasatinib, shown in Figure 3.40, which inhibits the function of a protein with high tyrosine kinase activity, leading to apoptosis, is indicated for the treatment of CML with Bcr-Abl translocation (Talpaz et al. 2018). This inhibitor acts at the ATP binding site on a tyrosine.

Nilotinib, as seen in Figure 3.40, is an inhibitor of the Bcr-Abl tyrosine kinase oncogene, resulting from a chromosomal translocation in a hematopoietic stem cell at the origin of CML and acute lymphoblastic leukemia (Philadelphia chromosome positive) (Jabbour and Kantarjian 2018). This property leads to its indication in the treatment of chronic myeloid leukemia with Bcr-Abl translocation.

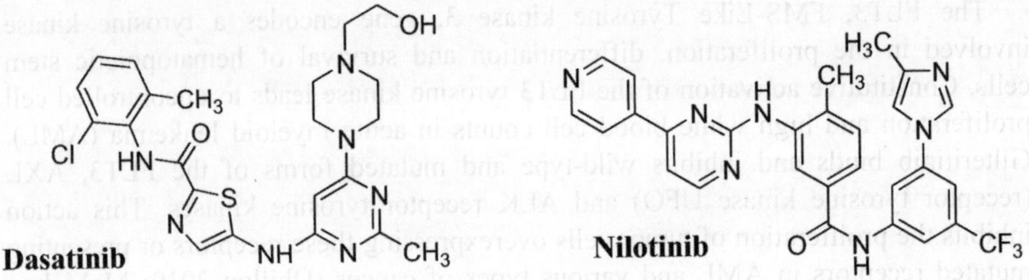

Figure 3.40. *Abl-Bcr kinase inhibitors*

B-Raf kinase

Vemurafenib (Figure 3.41) inhibits the serine–threonine kinase B-Raf, whose B-Raf V600 mutated form is implicated in melanoma development (Kim and Cohen 2016). V600 indicates that the "normal" valine at position 600 is substituted by another amino acid.

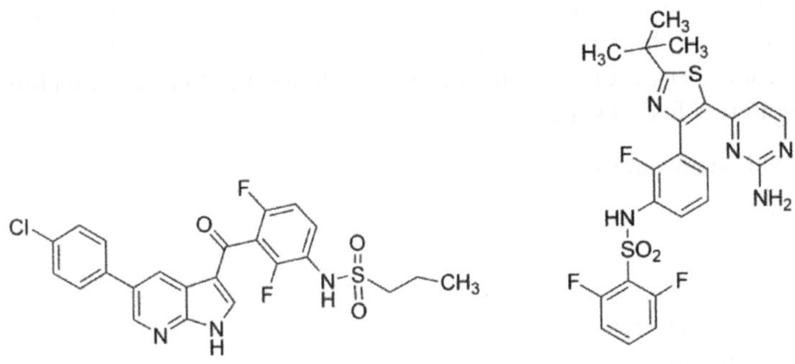

Figure 3.41. *B-Raf kinase inhibitors (normal and V600 mutated)*

Dabrafenib (Figure 3.41) binds selectively to the B-raf protein, leading to a decrease in MEK (MAPKinase Extracellular signal-regulated Kinases) and ERK (Extracellular signal-Regulated Kinases) phosphorylation, inducing the inhibition of tumor cells carrying the BRAF mutated gene (King et al. 2013). This molecule is indicated in cases of unresectable or metastatic melanoma carrying a B-Raf V600 mutation.

MEK kinases

Trametinib is a selective, reversible inhibitor of MEK1 and MEK2 exerting anti-proliferative activity against metastatic melanoma due to a BRAF V600 gene mutation (Lugowska et al. 2015).

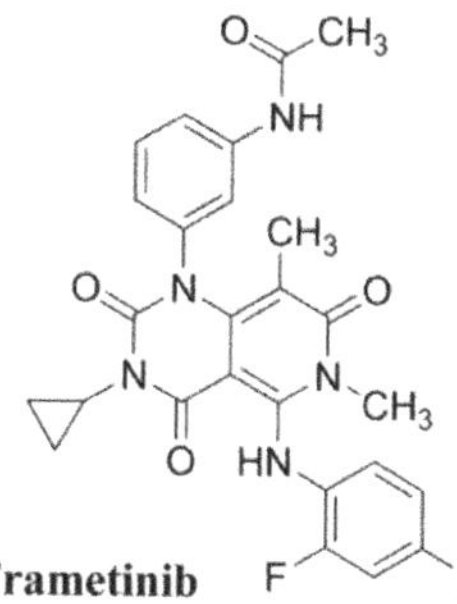

Trametinib

Trametinib is a highly selective, reversible, allosteric inhibitor of MEK1 and MEK2. MEK proteins are involved in signaling pathways regulated by ERK (extracellular signal-related kinase). Trametinib prevents MEK activation by B-Raf and inhibits MEK kinase activity. The compound is indicated in certain bronchial cancers or melanomas carrying the B-Raf V600 mutation (already seen in Vemurafenib and Dabrafenib).

BTK kinase

Ibrutinib is a potent, small molecule inhibitor of Bruton's tyrosine kinase (BTK). It is a substituted derivative of pyrazolo[3,4-d]pyrimidine.

It acts by binding and irreversibly inhibiting Bruton's tyrosine kinase activity. Its oral administration prevents activation and signaling mediated by B cells that are overexpressed in certain chronic lymphocytic leukemias and mantle cell lymphoma (Deeks 2017).

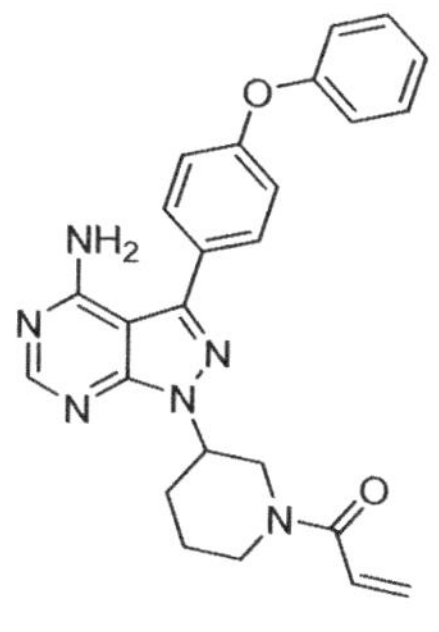

Ibrutinib

CDK 4/6 kinase

Ribociclib is an inhibitor of CDK 4/6 (kinases involved in cell cycle control). This drug is particularly indicated for the treatment of certain breast cancers (Edessa and Sisay 2017).

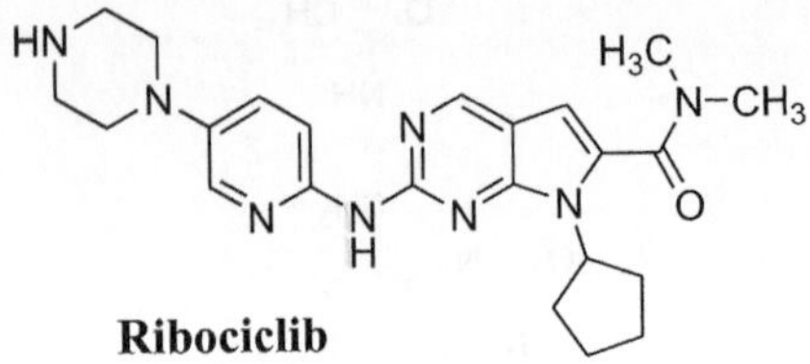

Ribociclib

PI3K-delta kinase

Idelalisib is a selective inhibitor of PI3K-delta (phosphoinositide-3-kinase delta). It is a substituted quinazoline derivative. It inhibits the activation of the phosphoinositide-3 kinase delta isoform (PI3K-delta) signaling pathway. This action inhibits tumor proliferation and induces apoptosis. This blocks the production of phosphatidyl-3,4,5-triphosphate, a second messenger, and the corresponding signaling process.

It is used in relapsed chronic lymphocytic leukemia in monotherapy (Shah and Mangaonkar 2015) or in combination with *Rituximab* (Furman et al. 2014).

The delta and gamma isoforms of phosphoinositide-3 kinase (PI3K) are overexpressed in hematological tumors and autoimmune inflammatory diseases. Research oriented towards the development of inhibitors has led to *Duvelisib* for its immunomodulatory and anti-proliferative properties (Blair 2018). This drug is used to treat chronic lymphoid leukemia, relapsed or refractory lymphocytic lymphoma after at least two previous treatments.

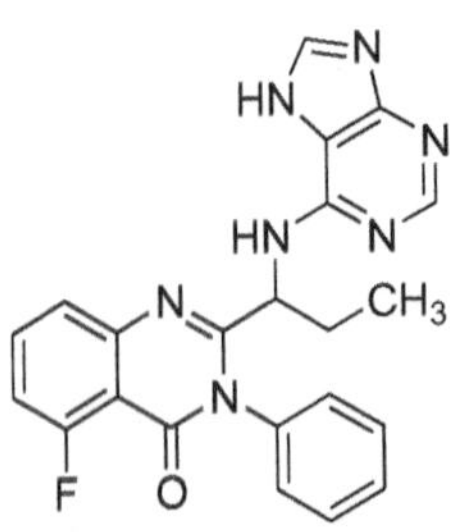

Idelalisib

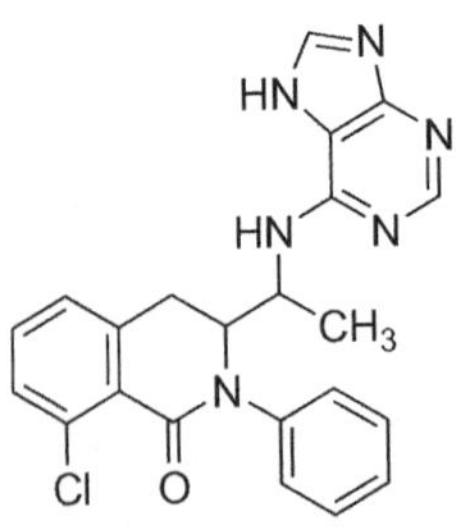

Duvelisib

mTor kinase

Temsirolimus is a selective inhibitor of the mTOR (Mammalian Target of Rapamycin) protein. mTOR is a serine–threonine kinase that is frequently deregulated in many human cancers (Porta et al. 2016). This drug is used in the treatment of certain kidney cancers. Temsirolimus is the ester of 2,2-bis-(hydroxymethyl)propionic acid and Sirolimus 1 (*Rapamycin*).

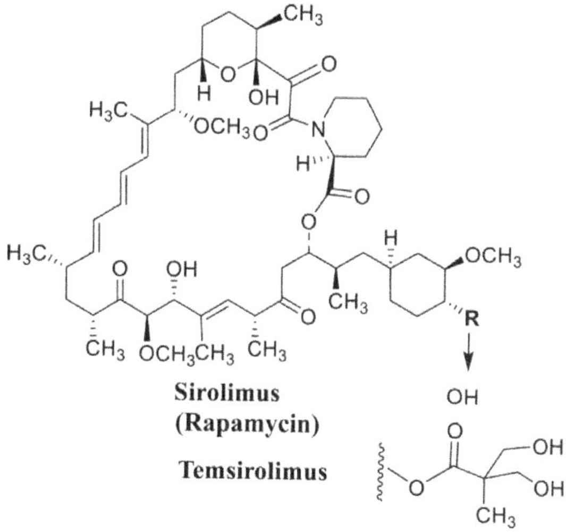

Sirolimus (Rapamycin)

Temsirolimus

C-mixed inhibitors (intracellular kinases and receptors)

One of the first kinase inhibitor derivatives is *Imatinib* (mesylate).

This drug acts primarily on Abl-Brc kinase, resulting in the following actions:

– competitive inhibition at the ATP-binding site of Abl-Brc protein tyrosine kinase;

– phosphorylation of protein tyrosine residues is blocked, preventing Abl-Brc-related signal transduction;

– the result is the inhibition of proliferation and induction of apoptosis in Ph chromosome-bearing cell lines.

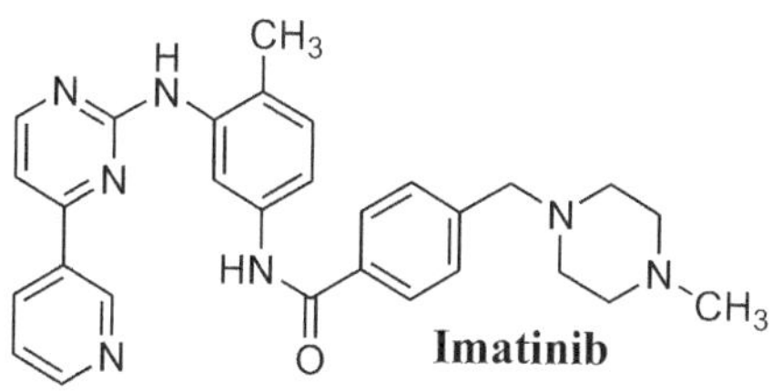

Imatinib

This drug is indicated for the treatment of Ph-positive chronic myeloid leukemia and acute lymphoblastic leukemia (Hirota et al. 1998). It inhibits other kinases. It also targets tyrosine kinase receptors such as the proto-oncogene c-Kit, CSF-1R (Colony Stimulating Factor receptor), and PDGF alpha and beta receptors. The substance selectively inhibits proliferation in Abl-Bcr tumor cell lines. The drug is used in the treatment of leukemias, myeloproliferative syndromes and certain gastrointestinal tumors.

Ponatinib is an acetylenic derivative of imidazo[1,2-b] pyridazine. Its action results in inhibitory activity against mutated Bcr-Abl and fibroblast growth factor receptors (FGFR) (Frankfurt and Licht 2013; Mingqiang et al. 2013). Ponatinib is an antineoplastic agent with applications in leukemias: chronic myeloid leukemia and Philadelphia chromosome-positive acute lymphoblastic leukemia. It carries the risk of adverse cardiovascular, cerebrovascular, peripheral vascular and venous thrombotic effects. Its prescription should be avoided in patients with a history of myocardial infarction or strokes, unless the expected benefit outweighs the risks.

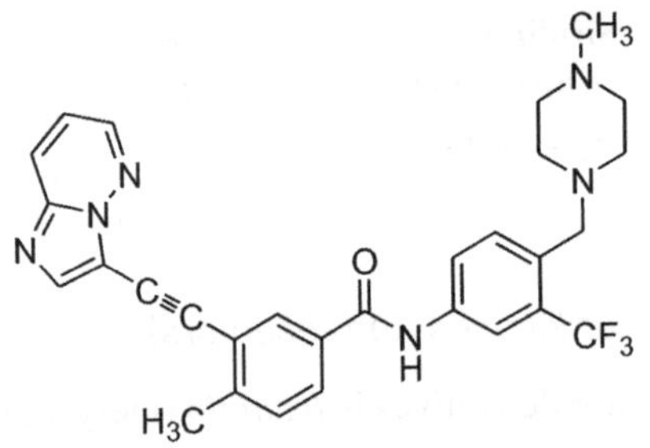

Ponatinib

Midostaurin inhibits a large number of receptor tyrosine kinases, such as FLT-3 and KIT (Abbas et al. 2019). This molecule blocks signaling pathways induced via these mutated or overexpressed receptors, leading to cell cycle arrest and apoptosis. It also targets PDGFR, VEGFR2 and members of the PKC (protein kinase C) family. This drug is indicated for acute myeloid leukemia.

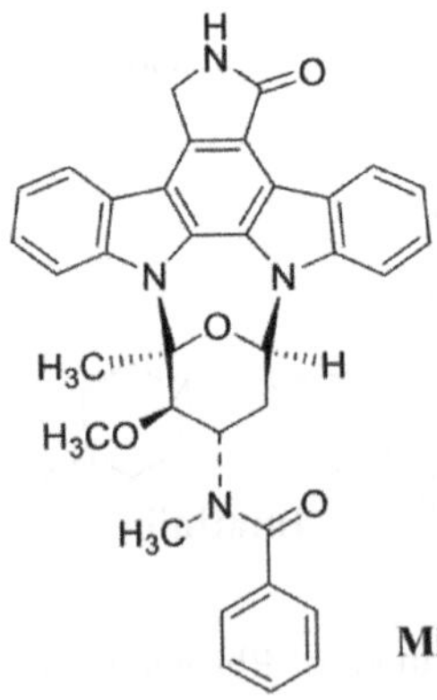

Midostaurin

Regorafenib, as well as two of its metabolites, M-2 and M-5, exhibit inhibitory activity against membrane and intracellular kinases: RET, VEGFR1, VEGFR2, VEGFR3, KIT, PDGFR-alpha, PDGFR-beta, FGFR1, FGFR2, TIE2, DDR2, TrkA, Eph2A, RAF-1, BRAF, BRAFV600, SAPK2, PTK5 and Abl. It is used to treat inoperable gastrointestinal tumors that are no longer responding to Imatinib or Sunitinib (Dhillon 2018; Heo and Syed 2018).

3.1.6.2.2. Other non-cancer pathologies – JAK kinases

Tofacitinib is a potent and selective inhibitor of the Janus Kinase (JAK) family (Dhillon 2017). In enzyme assays, this molecule acts on JAK1, JAK2 and JAK3, and to a lesser extent, on TyK2. JAK kinases are involved in cytokine receptor signaling pathways. Inhibition of these targets modulates inflammatory and immune responses. This molecule is used in the treatment of certain polyarthritis, rheumatic diseases and rectocolitis.

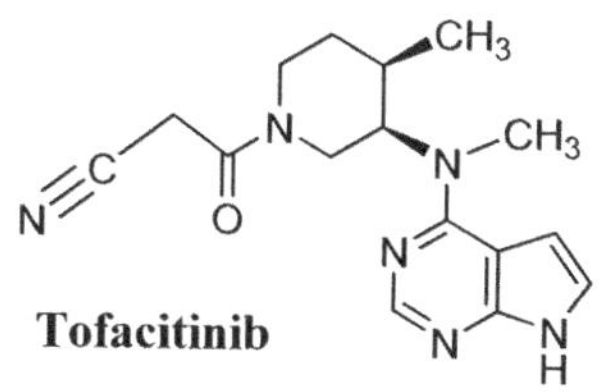

Tofacitinib

This latest drug completes the non-exhaustive list of molecules, used in therapeutics, that act on kinases.

Given the sheer number of research projects concerning this family of enzymes that are leading to the creation of new therapeutic molecules, they unfortunately cannot all be reported here, and have led us to make these choices.

3.2. Enzyme replacement therapy

3.2.1. *Gout*

Gout is a disease resulting from increased uricemia. Treatment involves the use of various hypouricemic and/or uricosuric drugs to reduce blood uric acid levels. It is also possible to use the action of uricase, an oxidase that is absent in humans. This enzyme is extracted from *Aspergillus flavus*. It catalyzes the conversion of uric acid into more soluble allantoin, facilitating its elimination. It can be used to treat primary or secondary hyperuricemia. A recombinant enzyme called *Pegloticase*, bonded to methoxypolyethylene glycol and produced by genetic engineering using a strain of *Escherichia coli*, has been developed for the treatment of this same pathology. However, the duration of these treatments is limited due to their immunogenicity.

3.2.2. *Acute leukemia and non-Hodgkin's lymphoma*

Another example is the use of *L-Asparaginase*, an enzyme that hydrolyzes L-asparagine (Asn) into L-aspartic acid (Asp), which is necessary for the growth of tumor cells that are unable to synthesize it. The absence of this amino acid blocks cell growth. Another example is *Crisantaspase*, asparaginase from *Erwinia chrysanthemi*, indicated for the treatment of acute lymphoblastic and myeloblastic leukemia, as well as non-Hodgkin's lymphoma in combination with other chemotherapeutic agents (Labrou et al. 2010; Salzer et al. 2014).

3.2.3. *Hypophosphatasia*

Asfotase alfa is a fusion protein that corresponds to the human alkaline phosphatase enzyme. It is designed to replace the non-specific alkaline phosphatase in defective tissue in a very rare genetic metabolic pathology known as hypophosphatasia. This orphan drug can be used to treat bone disorders related to this disease.

3.2.4. *Chronic obstructive pulmonary disease*

Alpha-antitrypsin (AAT) is a serine protease that inhibits elastase, a proteolytic enzyme. AAT deficiency gives rise to chronic obstructive pulmonary disease, such as emphysema, which frequently occurs in smokers (Ranes and Stoller 2005). It also develops in subjects with a genetic deficiency of this enzyme, leading to alterations in liver and skin function. Treatment consists of intravenous administration of transgenic or recombinant AAT (Respreeza, Prolastin, Alfalastin, Trypsone) (de Serres and Blanco 2014).

3.3. References

Abbas, H.A., Alfayez, M., Kadia, T., Ravandi-Kashani, F., Daver, N. (2019). Midostaurin in acute myeloid leukemia: An evidence-based review and patient selection. *Cancer Manag. Res.*, 11, 8817–8828.

Abdelgalil, A.A., Al-Kahtani, H.M., Al-Jenoobi, F.I. (2020). Erlotinib. *Profiles Drug Subst. Excip. Relat. Methodol.*, 45, 93–117.

Agarwal, V., Subash, A., Nayar, R.C., Rao, V. (2019). Is EGFR really a therapeutic target in head and neck cancers? *J. Surg. Oncol.*, 119(6), 685–686.

Arnoff, S.L., Berkowitz, K., Shreiner, B., Want, L. (2004). Glucose metabolism and regulation: Beyond insulin and glucagon. *Diabetes Spectrum*, 17, 183–185.

Balakumar, P., Nyo, Y.H., Renushia, R., Raaginey, D.R., Oh, A.N., Varatharajan, R., Dhanaraj, S.A. (2014). Classical and pleiotropic actions of dipyridamole: Not enough light to illuminate the dark tunnel? *Pharmacol. Res.*, 87, 144–150.

Barnes, P.J. (2013). Corticosteroid resistance in patients with asthma and chronic obstructive pulmonary disease. *J. Allergy Clin. Immunol.*, 131, 636–645.

Baumann, C.K. and Aebi, S. (2004). Aromatase inhibitors in the adjuvant therapy of breast carcinomas. *Ther. Umsch.*, 61, 359–364.

Bellesoeur, A., Carton, E., Alexandre, J., Goldwasser, F., Huillard, O. (2017). Axitinib in the treatment of renal cell carcinoma: Design, development, and place in therapy. *Drug Des. Devel. Ther.*, 11, 2801–2811.

Blair, H.A. (2018). Duvelisib: First global approval. *Drugs*, 78(17), 1847–1853.

Carswell, C.I., Polsker, G.L., Jarvis, B. (2002). Rosuvastatin. *Drugs*, 62, 2075–2085.

Carter, A.A., Gomes, T., Camacho, X., Juurlink, D.N., Shah, B.R., Mamdani, M.M. (2013). Risk of incident diabetes among patients treated with statins: Population based study. *BMJ*, 23, 346.

Chen, H., Jia, J., Ni, Z., Vastermark, A., Wu, B., Le, Y., Jawad, U. (2017). Orlistat response to missense mutations in lipoprotein lipase. *Biotechnol. Appl. Biochem.*, 64(4), 464–470.

Conti, M. and Beavo, J. (2007). Biochemistry and physiology of cyclic nucleotide phosphodiesterases: Essential components in cyclic nucleotide signaling. *Annu. Rev. Biochem.*, 76, 481–511.

Corbin, J.D., Francis, S.H., Webb, D.J. (2002). Phosphodiesterase type 5 as a pharmacologic target in erectile dysfunction. *Urology*, 60(2), 4–11.

Daugan, M., Noe, R., Fridman, W.F., Sautes-Fridman, C., Roumenina, L.T. (2017). Le système du complément. Une épée à double tranchant dans la progression tumorale. *Med. Sci. (Paris)*, 33(10), 871–877.

Dave, B., Gonzalez, D.D., Liu, Z.-B. (2016). Role of RPL39 in metaplastic breast cancer. *J Natl Cancer Inst.*, 109(6), 292.

Deeks, E.D. (2017). Ibrutinib: A review in chronic lymphocytic leukaemia. *Drugs*, 77(2), 225–236.

Dhillon, S. (2017). Tofacitinib: A review in rheumatoid arthritis. *Drugs*, 77(18), 1987–2001.

Dhillon, S. (2018). Regorafenib: A review in metastatic colorectal cancer. *Drugs*, 78(11), 1133–1144.

Dhillon, S. (2019). Gilteritinib: First global approval. *Drugs*, 79(3), 331–339.

DiNardo, C.D., Stein, E.M., de Botton, S., Roboz, G.J., Altman, J.K., Mims, A.M., Swords, R., Collins, R.H., Mannis, G.N., Pollyea, D.A. et al. (2018). Durable remissions with ivosidenib in IDH1-Mutated relapsed or refractory AML. *N. Engl. J. Med.*, 378(25), 2386–2398.

Doshi, A. and Chataway, J. (2017). Multiple sclerosis, a treatable disease. *Clin. Med. (Lond).*, 17(6), 530–536.

Edessa, D. and Sisay, M. (2017). Recent advances of cyclin-dependent kinases as potential therapeutic targets in HR+/HER2- metastatic breast cancer: A focus on ribociclib. *Cancer (Dove Med. Press)*, 9, 567–579.

Elkins, P., Coleman, D., Burgess, J., Gardner, M., Hines, J., Scott, B., Kroenke, M., Larson, J., Lightner, M., Turner, G. et al. (2014). Characterization of the isomeric configuration and impurities of (Z) endoxifen by 2D NMR, high resolution LC-MS, and quantitative HPLC analysis. *J. Pharm. Biomed. Anal.*, 88, 174–179.

Fiedler, L.R. (2018). Inhibiting the inhibitors, PTP1B as a therapeutic target in myocardial infarction. *Heart Res. Open J.*, 5(1), 8–11.

Forrest, R.R., Taylor, D.F., Fink, J.L., Gongora, M.M., Flegg, C., Teasdale, R.D., Suzuki, H., Kanamori, M., Kai, C., Hayashizaki, Y., Grimmond, S.M. (2006). PhosphoregDB: The tissue and sub-cellular distribution of mammalian protein kinases and phosphatases. *BMC Bioinforma.*, 7, 82.

Francis, S., Blount, M., Corbin, J. (2011). Mammalian cyclic nucleotide phosphodiesterases: Molecular mechanisms and physiological functions. *Physiol. Rev.*, 91, 651–690.

Frankfurt, O. and Licht, J.-D. (2013). Ponatinib – A step forward in overcoming resistance in chronic myeloid leukemia. *Clin. Cancer Res.*, 19(21), 5828–5834.

Furman, R.R., Jeff, P., Sharman, J.P., Coutre, S.E., Cheson, B.D., Pagel, J.M., Hillmen, P., Barrientos, J.C., Zelenetz, A.D., Kipps, T.J. et al. (2014). Idelalisib and rituximab in relapsed chronic lymphocytic leukemia. *N. Engl. J. Med.*, 370(11), 997–1007.

Giroux-Leprieur, E., Fallet, V., Cadranel, J., Wislez, M. (2016). Spotlight on crizotinib in the first-line treatment of ALK-positive advanced non-small-cell lung cancer: Patients selection and perspectives. *Lung Cancer (Auckl)*, 7, 83–90.

Goss, P.E. (2003). Breast cancer prevention–clinical trials strategies involving aromatase inhibitors. *J. Steroid Biochem. Mol. Biol.*, 86(3/5), 487–493.

Grande, E., Kreissl, M.C., Filetti, S., Newbold, K., Reinisch, W., Robert, C., Schlumberger, M., Tolstrup, L.K., Zamorano, J.L., Capdevila, J. (2013). Vandetanib in advanced medullary thyroid cancer: Review of adverse event management strategies. *Adv. Ther.*, 30(11), 945–966.

Gurdal, H., Tuglu, M.M., Bostanabad, S.Y., Dalkiliç, B. (2019). Partial agonistic effect of cetuximab on epidermal growth factor receptor and Src kinase activation in triple-negative breast cancer cell lines. *Int. J. Oncol.*, 54(4), 1345–1356.

Heo, Y.A. and Syed, Y.Y. (2018). Regorafenib: A review in hepatocellular carcinoma. *Drugs*, 78(9), 951–958.

Hirota, S., Isozaki, K., Moritama, Y., Hashimoto, K., Nishida T., Ishiguro, S., Kawano, K., Hanada, H., Kurata, A., Takeda, M. et al. (1998). Gain-of-function mutations of c-kit in human gastrointestinal stromal tumors. *Science*, 279, 577–580.

Jabbour, E. and Kantarjian, H. (2018). Chronic myeloid leukemia: 2018 update on diagnosis, therapy and monitoring. *Am. J. Hematol.*, 93(3), 442.

Kabakci, Z., Käppeli, S., Cantù, C., Jensen, L.D., König, C., Toggweiler, J., Gentili, C., Ribaudo, G., Zagotto, G., Basler, K. et al. (2019). Pharmacophore-guided discovery of CDC25 inhibitors causing cell cycle arrest and tumor regression. *Scientific Reports*, 9(1335) [Online]. Available at: https://www.nature.com/articles/s41598-019-38579-7.

Keravis, T. and Lugnier, C. (2012). Cyclic nucleotide phosphodiesterase (PDE) isozymes as targets of the intracellular signalling network: Benefits of PDE inhibitors in various diseases and perspectives for future therapeutic developments. *Br. J. Pharmacol.*, 165, 1288–1305.

Khotskaya, Y.B., Holla, V.R., Farago, A.F., Mills Shaw, K.R., Meric-Bernstam, F., Hong, D.S. (2017). Targeting TRK family proteins in cancer. *Pharmacol. Ther.*, 173, 58–66.

Kim, E.S. (2017). Enasidenib: First global approval. *Drugs*, 77(15), 1705–1717.

Kim, A. and Cohen, M.S. (2016). The discovery of vemurafenib for the treatment of BRAF-mutated metastatic melanoma. *Expert Opin. Drug Discov.*, 11(9), 907–916.

Kim, D.W., Talati, C., Kim, R.J. (2017). Hepatocellular carcinoma (HCC): Beyond sorafenib chemotherapy. *Gastrointest. Oncol.*, 8(2), 256–265.

King, A.J., Arnone, M.R., Bleam, M.R., Moss, K.G., Yang, J., Fedorowicz, K.E., Smitheman, K.N., Erhardt, J.A., Hughes-Earle, A., Kane-Carson, L.S. et al. (2013). Dabrafenib; preclinical characterization, increased efficacy when combined with trametinib, while BRAF/MEK tool combination reduced skin lesions. *PLoS One*, 8(7), e67583.

Korntnera, S., Pomorskib, A., Krelb, A., Bishopa, A.C. (2018). Optimized allosteric inhibition of engineered protein tyrosine phosphatases with an expanded palette of biarsenical small molecules. *Bioorg. Med. Chem.*, 26(9), 2610–2620.

Kosterlitz, H. and Hughes, W. (1977). Opiate receptors and endogenous opioid peptides in tolerance and dependence. *J. Adv. Exp. Med. Biol.*, 85, 141–154.

Kurzrock, R., Sherman, S.I., Ball, D.W., Forestiere, A.A., Cohen, R.B., Mehra, R., Pfister, D.G., Cohen, E.E., Janisch, L., Nauling, F. et al. (2011). Activity of XL184 (Cabozantinib), an oral tyrosine kinase inhibitor, in patients with medullary thyroid cancer. *Clin Oncol.*, 29(19), 2660–2666.

Labrou, N.E., Papageorgiou, A.C., Avramis, V.I. (2010). Structure-function relationships and clinical applications of L- asparaginases. *Curr. Med. Chem.*, 17, 2183–2195.

Ledermann, J.A., Oza, A.M., Lorusso, D., Aghajanian, C., Oaknin, A., Dean, A., Colombo, N., Weberpals, J.I., Clamp, A.R., Sdambia, G. et al. (2020). Rucaparib for patients with platinum-sensitive, recurrent ovarian carcinoma (ARIEL3): Post-progression outcomes and updated safety results from a randomised, placebo-controlled, phase 3 trial. *Lancet Oncol.*, 21(5), 710–722.

Llorens, L., Gacel, G., Swerts, J.P., Perdisot, R., Fournier-Zaluski, M.C., Schwartz, J.C., Roques, B.P. (1980). Rational design of enkephalinase inhibitors: Substrate specificity of enkephalinase studied from inhibitory potency of various dipeptides. *Biochem. Biophys. Res. Commun.*, 96, 1710.

Lugowska, I., Kosela-Paterczyk, H., Kozak, K., Rutkowski, P. (2015). Trametinib: A MEK inhibitor for management of metastatic melanoma. *Onco Targets Ther.*, 8, 2251–2259.

Massarelli, E. and Papadimitrakopoulou, V. (2015). Ceritinib for the treatment of late-stage (metastatic) non-small cell lung cancer. *Clin. Cancer Res.*, 21(4), 670–674.

McMahon, C.M. and Perl, A.E. (2019). Gilteritinib for the treatment of relapsed and/or refractory FLT3-mutated acute myeloid leukemia. *Expert Rev. Clin. Pharmacol.*, 12(9), 841–849.

Mingqiang, R., Haiyan, Q., Cowell, J.K. (2013). Ponatinib suppresses the development of myeloid and lymphoid malignancies associated with FGFR1 abnormalities. *Leukemia*, 27(1), 32–40.

Mishanina, T.V., Koehn, E.M., Kohen, A. (2012). Mechanisms and inhibition of uracil methylating enzymes. *Bioorg. Chem.*, 43, 37–43.

Mouridsen, H., Gershanovich, M., Sun, Y., Pérez- Carrion, R., Boni, C., Monnier, A., Apffelsteadt, J., Smith, R., Sleeboom, H.P., Jänicke, F. et al. (2001). Superior efficacy of letrozole versus tamoxifen as first- line therapy for postmenopausal women with advanced breast cancer: Results of a phase III study of the international letrozole breast cancer group. *J. Clin. Oncol.*, 19, 2596–2606.

Muchowski, J.M., Unger, S.H., Ackrell, J., Cheung, P., Cooper, G.F., Cook, J., Gallegra, P., Halpern, O., Koehler, R., Kluge, A.F. et al. (1985). Synthesis and antiinflammatory and analgesic activity of 5-aroyl-1,2-dihydro-3H-pyrrolo[1,2-a]pyrrole-1-carboxylic acids and related compounds. *J. Med. Chem.*, 28, 1037–1049.

Nishimura, M., Takahashi, H., Yoshimur, M. (2000). Role of nitric oxide in haemodialysis hypotension. *Acta. Physiol. Scand.*, 168(1), 181–186.

Nolting, M., Schneider-Merck, T., Trepel, M. (2014). Lapatinib. *Recent Results Cancer Res.*, 201, 125–143.

Ozdemir, O., Boran, M., Gokçe, V., Uzun, Y., Koçak, B., Korkmaz, S. (2000). A case with severe rhabdomyolysis and renal failure associated with cerivastatin-gemfibrozil combination therapy – a case report. *Angiology*, 51, 695–697.

Padala, S. and Thompson, P.D. (2012). Statins as a possible cause of inflammatory and necrotizing myopathies. *Atherosclerosis*, 222(1), 15–21.

Pagidipati, N.J. and Gaziano, T.A. (2013). Estimating deaths from cardiovascular disease: A review of global methodologies of mortality measurement. *Circulation*, 127, 749–756.

Porta, C., Tortora, G., Larkin, J.M., Hutson, T.E. (2016). Management of poor-risk metastatic renal cell carcinoma: Current approaches, the role of temsirolimus and future directions. *Future Oncol.*, 12(4), 533–549.

Quan, C., Xiao, J., Liu, L., Duan, Q., Yuan, P., Zhu, F. (2017). Protein kinases as tumor biomarkers and therapeutic targets. *Curr. Pharm. Des.*, 23(29), 4209–4225.

Ranes, J. and Stoller, J.K. (2005). A review of alpha-1 antitrypsin deficiency. *Semin. Respir. Crit. Care Med.*, 26(2), 154–166.

Rizzo, M. and Porta, C. (2017). Sunitinib in the treatment of renal cell carcinoma: An update on recent evidence. *Ther. Adv. Urol.*, 9(8), 195–207.

Robinson, D.R., Wu, Y.M., Lin, S.F. (2000). The protein tyrosine kinase family of the human genome. *Oncogene*, 19, 5548–5557.

Salzer, W.L., Asselin, B.L., Plourde, V., Corn, T., Hunger, S.P. (2014). Development of asparaginase *Erwinia chrysanthemi* for the treatment of acute lymphoblastic leukemia. *Ann N Y Acad. Sci.*, 1329, 81–92.

Sarkar, S.R., Kaitwatcharachai, C., Levin, N.W. (2004). Nitric oxide and hemodialysis. *Semin. Dial.*, 17(3), 224–228.

Sarosiek, T. and Morawski, P. (2018). Trastuzumab and its biosimilars. *Pol. Merkur. Lekarski*, 44(263), 253–257.

Schudt, C., Hatzelmann, A., Beume, R., Tenor, H. (2011). Phosphodiesterase inhibitors: History of pharmacology. *Handb. Exp. Pharmacol.*, 204, 1–46.

de Serres, F. and Blanco, I. (2014). Role of alpha-1 antitrypsinin human health and disease. *J. Intern. Med.*, 276(4), 311–335.

Shah, A. and Mangaonkar, A. (2015). Idelalisib: A novel PI3K inhibitor for chronic lymphocytic leukemia. *Ann. Pharmacother.*, 49(10), 1162–1170.

Sharma, R., Joubert, J., Malan, S.F. (2018). Recent developments in drug design of NO-donor hybrid compounds. *Mini Rev. Med. Chem.*, 18(14), 1175–1198.

Smyth, D.G. (2016). 60 YEARS OF POMC: Lipotropin and beta-endorphin: A perspective. *J. Mol. Endocrinol.*, 56(4), T13–T25.

Stuehr, D.J. and Mahfuzul Haque, M.M. (2019). Nitric oxide synthase enzymology in the 20 years after the Nobel Prize. *Br. J. Pharmacol.*, 176(2), 177–188.

Talley, J.J., Bertenshaw, S.R., Brown, D.L., Carter, J.S., Graneto, M.J., Kellogg, M.S., Koboldt, C.M., Yuan, J., Zhang, Y.Y., Seibert, K. (2000). N-[[(5-methyl-3-phenylisoxazol-4yl)- phenyl] sulfonyl]propanamide, sodium salt, parecoxib sodium: A potent and selective inhibitor of COX-2 for parenteral administration. *J. Med. Chem.*, 43, 1661–1663.

Talpaz, M., Saglio, G., Atallah, E., Rousselot, P. (2018). Dasatinib dose management for the treatment of chronic myeloid leukemia. *Cancer*, 124(8), 1660–1672.

Tarkin, J.M. and Kaski, J.C. (2016). Vasodilator therapy: Nitrates and nicorandil. *Cardiovasc. Drugs Ther.*, 30(4), 367–378.

Wakeling, A.E., Guy, S.P., Woodburn, J.R., Ashton, S.E., Curry, B.J., Barker, A.J., Gibson, K.H. (2002). ZD1839 (Iressa): An orally active inhibitor of epidermal growth factor signaling with potential for cancer therapy. *Cancer Res.*, 62, 5749.

Xu, B., Konze, K.D., Jin, J., Wang, G.G. (2015). Targeting EZH2 and PRC2 dependence as novel anticancer therapy. *Exp. Hematol.*, 43(8), 698–712.

Yan, K.S., Yan-Kin, C., Loao, T.W., Peng C.-M., Lee, S.-C., Liu, Y.-J., Chan, W.P., Chou, R.-H. (2017). EZH2 in cancer progression and potential application in cancer therapy: A friend or foe. *Int. J. Mol. Sci.*, 18(6), 1172.

Zhang, K.Y., Ibrahim, P.N., Gillette, S., Bollag, G. (2005). Phosphodiesterase-4 as a potential drug target. *Expert Opin. Ther. Targets*, 9(6), 1283–1305.

4

What are the New Targets?

4.1. Inhibition and activation strategies: advantages, disadvantages and current status

Currently, for some enzyme-targeting drug candidates (scientific publications), there is a lack of accuracy on their activation or inhibition characteristics from a kinetic point of view.

At the pharmaceutical level, all ADME (Administration, Distribution, Metabolism and Excretion) Tox in vitro and in vivo tests are carried out, and the product's efficacy on the target is always solidly proven. A plethora of experiments have been carried out, including IC_{50} calculations, binding experiments, surface plasmon resonance, crystallography coupled with X-ray diffraction, nuclear magnetic resonance and molecular modeling. However, the kinetic foundations of enzymology tend to be neglected in this frantic race.

Characterizing the inhibition or activation exerted by the product through the development of kinetic protocol(s) that are specific to each enzyme is no longer always done. Yet, it would enable us to fully understand the intimate relationship between the two factors. It is not just a question of stating that the two molecules are linked, but what the consequences of this affinity are, for the way the system works. Putting a useful past behind us is not the best way to prepare for a serene future.

Some of the pharmaceutical industry's setbacks can be linked to an eagerness to discover, without taking all the necessary precautions. The mindset of one of the most valiant scientists, Louis Pasteur, shows the path that we should never lose sight of. He exhausted himself chasing down microbes, overcoming many of the diseases that have decimated so many human beings. In his fight against rabies, he went through complicated times, when his vaccination protocol was called into question. However, his rigorous experimental protocols and the dedication of his entire team enabled him to achieve his noble goal: to roll back infections for humankind. He finally concluded, at the end of a busy scientific life: "When you reach the final goal, you have to be able to say to yourself, I did what I could!" This sentence was addressed to the Sorbonne students who came to hear him speak for the last time in 1892.

More and more robust, ecological, imaginative, economical and easy-to-use enzymatic tests are being designed by both private and public laboratories. They must be used as sparingly as possible, so that they provide all the necessary information.

The advantages and disadvantages of the various effectors acting on enzymes, in order to emphasize the importance of determining their mechanism of action will now be reviewed.

4.1.1. *Irreversible inhibitors*

These inhibitors are characterized by the fact that they bind covalently to the target enzyme. Dialysis or simple washing of the complex enzyme will not dissociate the entity formed. Irreversible activators could be considered, but there are no well-known examples.

A covalent modification is often necessary when irreversible inhibitors or activators are involved.

An inhibitor metabolized by the enzyme, which permanently inactivates the catalytic site or a substrate binding site, is termed a suicide substrate. The enzymatic reaction can no longer be catalyzed.

Kinetic characterization of an irreversible inhibitor can be found in the first chapter of this book.

4.1.1.1. *Advantages*

Many of these inhibitors are used by substrates. They are modified to bind the enzyme irreversibly. In many cases, this involves chemical work to modify existing structures or create new ones based on data that is already known. The databases (bibliography, as well as sitography, thanks to the Internet) that are available for such work are a positive point. These inhibitors can be used to target areas of enzymes other than the active sites, in order to inactivate them. This is a possible broadening of applications. While these inhibitors are of pharmaceutical interest, they are also valuable for fundamental understanding of enzyme function. For example, cytochrome p450 suicide inhibitors can either prevent substrate entry into the active site, disrupt electron transfer, facilitate heme dissociation or destructure the protein, leading to its degradation. They have been instrumental in understanding how this family of enzymes works. In therapeutic terms, Exemestane is an irreversible steroidal inhibitor of aromatase (an enzyme complex with a cytochrome p450 unit), structurally related to the enzyme's natural substrate, androstenedione. It is used effectively in hormone-dependent breast cancer in post-menopausal women. Many irreversible inhibitors were developed or are under development.

4.1.1.2. *Disadvantages*

The major disadvantage of treatments based on such molecules is that the process is irreversible. There is no immediate return once the treatment has been administered. It is necessary to wait for the inhibitor to degrade and for the enzyme to be produced. The covalent link prevents any dosing errors or poor targeting.

This problem is often coupled with the molecule's insufficient initial specificity of action. This needs to be improved by making the compound highly specific to the target (QSAR studies, molecular modeling, Schechter and Berger (1967) nomenclature studies).

Molecules in this category have been used as "weapons of war", resulting in massive crimes against living organisms, and therefore against humanity. Organophosphorus compounds targeting acetylcholinesterase by covalently phosphorylating the active site cause muscle paralysis, convulsions, bronchoconstriction and ultimately death by asphyxiation in mammals, including human beings.

4.1.2. *Competitive inhibitors*

Precise knowledge of an enzyme's tertiary structure enables us to tailor the inhibitor to the enzyme. The active site is very often the best-known and most studied area of an enzyme. Competitive inhibitors are the most extensively studied and developed. The kinetic characteristics of a competitive inhibitor in a Michaelian system were presented in the first chapter of this book.

4.1.2.1. *Advantages*

In the case of a competitive inhibitor whose over-administration may have caused a problem, it may be possible to lift the inhibition by adding substrate at a saturated dose. Competitive inhibitors are very often used by the structure of enzymatic substrates, either directly or indirectly, through the reaction intermediates obtained during the enzymatic reaction. This simplifies molecular design. The substrate, the active site domain, the catalytic sequence and the significant areas for establishing weak bonds are all known. For this reason, a very large number of enzyme-targeting drugs are competitive inhibitors. It should be noted that inhibition characterization kinetics are not always performed. However, they are the only way of unambiguously characterizing the mode of action of an effector molecule.

Knowledge of a large number of active enzyme sites also reduces the risk of undesirable effects on related enzymes (comparison principle thanks to enzymatic classification in particular, and the numerous feedbacks from experience with related molecules).

4.1.2.2. *Disadvantages*

The major disadvantage is directly related to the main advantage of this family of inhibitors. Indeed, developing this family of inhibitors on a very large scale logically leads to shortcomings in work on other effectors (financial deficiencies, human resources for experiments and other avenues of research).

Working predominantly on a single type of inhibitor to prevent the development of pathogenic microorganisms is not the best way to counter resistance. Resistance arises from point mutations in the primary structure of an enzyme. These resistances can be overcome by inhibitors from the same family. However, having several inhibitors from different families targeting different domains of the enzyme is a more effective solution. This provides a wider choice of treatments, and just as many solutions for controlling mutants. The problem of resistance is added to that of side effects. These affect all drugs. Although inhibitors from the same family can present different side effects, having different and varied families of molecules once again leads to a diversification of treatment choices. This makes it possible to offer several solutions, so that the patient can cope as well as possible with the care provided.

Research cannot solely focus on irreversible and competitive inhibitors. It is necessary to find a new balance that makes room for other types of effectors, without thinking about profitability in the health sector. Life remains our most precious asset. It is a difficult path to tread, one that requires constant reassessment of the evolution of the benefit/risk ratio, both for the patients treated and for the pharmaceutical industry, which generates new, very costly solutions. Treatments can thus evolve to better combat pathologies, taking into account the active ingredients and the administered doses.

Enzyme effectors behind irreversible or competitive inhibitors are considered below.

4.1.3. *Non-competitive inhibitors*

The study of an enzyme's tertiary structure should not essentially be limited to that of its active site. Searches for other areas with sufficiently characteristic spatial structures (pockets, tunnels, grooves) should be systematic. Non-competitive inhibitors are less numerous, but could potentially be more important if research fields were broadened. These inhibitors can bind to the free enzyme or to the complex with the same or different affinities. The result is another binding site, different from the substrate binding site. The binding of the inhibitor molecule to the enzyme prevents it from changing conformation to transform the substrate correctly, if the latter is bound. If the substrate is not bound, the conformational change brought about by the inhibitor may modify the binding and subsequently disrupt the transformation of the substrate. Sometimes, the non-competitive inhibitor is assimilated to an allosteric inhibitor. Limiting our notion of other sites, this description is not misleading. However, it does not consider the notion of cooperativity that is inherent in multimeric enzymes and allosteric effectors. The use of the term *allosteric* thus becomes irrelevant. For this reason, non-competitive inhibitors and allosteric inhibitors have been treated separately in the first chapter of the book.

The kinetic characteristics of a non-competitive inhibitor in a Michaelian system have been presented in the previous chapters.

4.1.3.1. *Advantages*

If the organism tries to compensate for the inhibition by making more substrate available to the enzyme, there will be no effect, as the inhibitor's binding site is different from that of the substrate. Inhibition cannot therefore be lifted by excess substrate. If the dosage is properly calibrated, the inhibitor's action will not be disrupted by an increase in substrate concentration.

Non-competitive inhibitors are among the effectors to be developed to diversify the families of molecules targeting enzymes. There are many more avenues of research. At present, few are used therapeutically. Examples include Acetazolamide, a non-competitive inhibitor of carbonic anhydrase in relation to its substrate, carbon dioxide and non-nucleoside inhibitors of HIV reverse transcriptase in relation to the enzyme's nucleotide substrates.

Their development on a larger scale would enable us to gain a better understanding of how enzymes work. There is also a fundamental interest.

4.1.3.2. *Disadvantages*

If the inhibitor is not used under optimum conditions (such as excessive dosage), the competitive removal of inhibition by the enzyme substrate will not be possible.

Developing inhibitors like this requires greater financial and human resources. Studies and bibliographical contributions are less significant than for competitive inhibitors. They have to be compensated for by more preliminary research work.

4.1.4. *Uncompetitive inhibitors*

The problems that uncompetitive inhibitors have are the same as those described for non-competitive inhibitors, given the very limited number of effectors described of this type. These inhibitors only bind to the substrate/enzyme complex. As with non-competitive inhibitors, this results in a different binding site to that of the substrate. The binding of the molecule to the enzyme/substrate complex prevents proper substrate transformation. Unlike non-competitive and competitive inhibitors, the enzyme alone is not the target of the inhibitor. It should be pointed out that this type of inhibitor would be a very interesting model for the enzymatic interactome, or *enzymatome*. Potential binding sites could be revealed, not by the binding of the substrate to the enzyme, but by the binding of another cellular partner. This relationship would lead to conformational changes in the enzyme, which could result in the appearance of transient domains targeted by effectors.

The kinetic characteristics of an uncompetitive inhibitor in a Michaelian system were presented at the start of the book.

4.1.5. *Allosteric effectors*

Inhibitors, or allosteric activators, are positioned at sites that are separate from the substrate binding site. These allosteric domains modulate the activity of multimeric enzymes (made up of several sub-units). They are capable of generating cooperativity between protein monomers. Binding of the effector to one sub-unit modifies the behavior of neighboring sub-units. These enzymes are frequently involved in metabolism. The kinetic characteristics of allosteric effectors are presented at the beginning of the book.

4.1.5.1. *Advantages*

Numerous allosteric enzymes and natural effectors targeting them are described in the literature. The structures, modes of action and interactions of these effectors are the subject of numerous studies, providing a solid foundation. These effectors are regulators. Precise dosing according to pathologies and patients could, in the long-term, enable a very fine-tuned regulation of enzyme activity.

4.1.5.2. *Disadvantages*

As there are very often natural allosteric inhibitors and activators for a given system, it is necessary to develop an inhibitor or activator while ensuring that its action is not offset by its natural opposite. Moreover, as lots of natural allosteric effectors can act upon numerous enzymes, it is necessary to check the specificity of a synthetic effector by testing it on every possible target. Developing such inhibitors also requires more financial and human resources than for competitive inhibitors, as other issues need to be resolved.

4.1.6. *Other strategies*

There are other inhibition and activation strategies:

– Bifunctional effectors, to act on two targets of the same pathogen.

– An enzyme catalyzing a chemical reaction is very rarely alone. It is often part of an evolving molecular complex. By taking into account the notion of the interactome and drawing inspiration from uncompetitive inhibitors, knowledge of the links between molecular partners and their consequences on the spatial structure of the enzyme would make it possible to discover transient sites at which specific effectors would act.

– Oligomerization inhibitors also represent an interesting avenue for development. In the living world, there are numerous multimeric enzymes. We have seen allosteric enzymes, but they are not the only ones. The aim is to generate molecules that prevent the association or the dissociation of complexes of such enzymes, so as to prevent their activity.

These last three concepts can be applied to other proteins, or even other molecules.

A number of publications describe such effectors. However, few therapeutic applications are currently available.

4.2. Exogenous and endogenous targets

These examples are not exhaustive. There are a considerable number of enzymes involved in vital processes. In humans, it ranges from 2,000 to 5,000 enzymes, depending on the isoforms. Consequently, targeting these enzymes (inhibition, activation or replacement) represents a great potential for therapeutic applications. Therefore, we have limited ourselves to giving a few examples of new endogenous and exogenous targets to be targeted, or already-known targets to be targeted differently.

4.2.1. *Exogenous targets*

For pathogenic microorganisms in general, there is an emphasis on the search for reliable and effective vaccine protocols that prevent the onset of infectious pathologies. In many countries, there is an outcry against vaccines. It is imperative to recall the historical and scientific health benefits of vaccines, and the objectives for future improvements. These transparent reminders are essential for the entire population, especially the very young.

When vaccine research fails to provide an effective solution, there is the option of drug treatment, either by improving active ingredients or by finding new targets. Sections 4.2.1.1–4.2.1.9 look at new avenues of research in enzymology that could lead to new therapeutic applications.

4.2.1.1. *Coronavirus; SARS-CoV-2: RNA-dependent RNA polymerase (EC 2) and 3C like-protease (EC 3)*

Coronaviruses (CoV) belong to a large family of RNA viruses, the *Coronaviridae*. They are defined by a crown-shaped capsule, hence their name. Many sub-types infect different species. At least five can infect humans, of which HCoV-229 and HCoV-OC43 are the most common.

Widespread, these viruses are often associated with colds and mild flu syndromes. However, three of them – SARS-CoV, MERS-CoV and SARS-CoV – have caused major public health problems over the last 20 years. The year 2020 was marked by the most serious pandemic linked to SARS-CoV-2.

The RNA-positive SARS-CoV-2 virus is a coronavirus responsible for the severe acute respiratory syndrome known as Covid-19. Infection takes place mainly via the respiratory tract, and after attachment to the host cell, the virus enters in and multiplies. An essential step in the viral multiplication cycle is catalyzed by the RNA-dependent RNA polymerase (RdRP) enzyme complex (Gao et al. 2020). Positive-strand viral genomic RNA is transcribed into a complementary negative-strand RNA, which is used as a template for the synthesis of positive-strand viral mRNA and genomic RNA. Blocking this viral enzyme is an effective target for inhibiting viral replication.

In parallel, another viral enzyme, a protease, is targeted. This is the 3C-like protease.

To date, *Molnupiravir* (MK-4482), a replication-disrupting ribonucleotide analogue, has been shown to be effective in combating the virus. However, this orally administered drug is not recognized as being sufficiently effective by all national or supranational drug agencies.

Therapies based on the use of monoclonal antibodies are available to combat the virus (*Casirivimab/Imdevimab* – or *Bamlavinimab/Etesevimab*).

Since the end of 2020, vaccines have successfully passed clinical phases and have been dispensed to the global population (mainly in developed countries). Publications reporting the results of clinical trials are available. Readers interested in this field can consult recent reviews of the numerous research projects underway to develop vaccines (Liu et al. 2020), as well as various substances and drugs already known to have potential activity against this virus (Zhou et al. 2020). Trials are still underway, and to date, very positive results validate certain mRNA-based antigenic vaccines and protein vaccines, as well as the start of antiviral biotherapies and chemotherapies.

The example of the SARS-CoV-2 pandemic is an interesting one to analyze. It shows that, in the absence of long-term, financially sustainable research, our societies have to proceed on the basis of absolute urgency, with no leeway in the face of such epidemics. In the case of emerging, potentially dangerous pandemics, it is necessary to test effective drugs on other targets and, if possible, envisage new vaccine protocols under very delicate conditions.

4.2.1.2. *HIV protease (EC 3)*

For the HIV protease (EC 3), targeting the zone that is mainly involved in its dimerization, the antiparallel beta-sheet, would be of interest (Schramm et al. 1992; Franciskovich et al. 1993; Dumond et al. 2003). These are oligomerization inhibitors (dimerization, in this case). This zone is found to be little mutated when antiprotease-resistant mutants are studied. This is a very interesting point for therapeutic molecules.

4.2.1.3. *Integrase of the same virus*

Integrase from the same virus is an enzyme that functions in oligomeric form. Inhibitors targeting the domains involved in oligomerization are among the interesting leads to be explored (Camarasa et al. 2006).

4.2.1.4. *HIV reverse transcriptase (EC 2)*

For HIV – RT reverse transcriptase (EC 2), there are numerous other inhibition strategies:

– NNRTIs can bind to the RT allosteric pocket, while disrupting the active site with an aliphatic chain grafted to the inhibitor molecule. This type of molecule could disrupt polymerase function, while attached to the pocket (Tronchet et al. 2002; Dumond et al. 2020).

– Inhibitors with two viral targets, including RT, referred to as bifunctional inhibitors, can be considered (Stevens et al. 2003).

– Inhibiting the enzyme's RNase H activity may also be considered (Yu et al. 2008).

– As RT is a dimeric enzyme, dimerization inhibitors are possible (Divita et al. 1995).

4.2.1.5. *Respiratory syncytial virus (RSV)*

Virus-dependent RNA polymerase (EC 2) is a highly attractive potential therapeutic target. RSV is the main cause of severe respiratory illnesses in young children (70% of bronchiolitis cases). Vaccines and antivirals are limited. RSV is an enveloped, negative-polarity, single-stranded RNA virus transcribed and replicated by the viral RNA-dependent RNA polymerase (RdRP) complex.

Currently, there is no validated inhibitor of the viral enzyme for therapeutic use. The only treatment approved is *Palivizumab*, a 95% humanized IgG1K (Immunoglobulin G1K) monoclonal antibody. It is oriented against the antigenic site A of the RSV fusion protein.

4.2.1.6. *Dihydrodipicolinate synthase (DHPDS) (EC 4)*

This tetrameric enzyme, required for the biosynthesis of bacterial peptidoglycan, catalyzes the condensation of pyruvate on semialdehyde L-aspartate. This association forms a cyclic intermediate which, following dehydration, yields dihydrodipicolinate. For several years now, DHDPS has been considered a target of interest for the development of new antibiotics. An interesting first strategy was led by a Canadian team to create the first allosteric inhibitor of this catalyst. This new inhibitor binds to the lysine-binding allosteric pocket. This effector, bislysine, consists of two lysines linked by an ethylene bridge. The inhibitor binds to all forms of the enzyme, maintaining a bond between the enzyme and its two substrates. The molecule is 200–300 times more refined than the lysine. Hill's representations show the cooperative nature of inhibition. A very interesting point shows that the enzyme mutant Y110F, resistant to lysine inhibition, is strongly inhibited by this new inhibitor (Skovpen et al. 2016). As this enzyme is tetrameric, tetramerization inhibitors could also be considered (Perugini et al. 2005).

4.2.1.7. *GDP-mannose pyrophosphorylase (GDP-MP) (EC 2)*

This enzyme is characteristic of some 20 protozoa of the *Leishmania* genus, which, by infecting mammals such as humans, give rise to leishmaniasis. These tropical diseases are now found in 88 countries worldwide. Several hundred thousand people are infected every year.

In the case of *Leishmania mexicana*, the enzyme is active in the form of a hexamer. One of the main avenues of research to counter these parasites is to develop new molecules that alter the biosynthesis of glycoconjugates, which are essential for microorganisms to be recognized by macrophages and subsequently internalized. Multimeric GDP-MP is one of the enzymes that are essential to this glycoconjugate synthesis pathway. A classic route to pathogen inactivation is to develop the usual competitive inhibitors (Pomel et al. 2019). Another route is the development of enzyme oligomerization inhibitors (Perugini et al. 2005).

4.2.1.8. *Fatty acid synthase (EC 2)*

This enzyme complex is present in many living species in various forms. The mycobacterial type I fatty acid synthase complex is especially targeted in therapeutics. It is required for the elaboration of fatty acid chains. It is a homohexamer (six sub-units). Cerulenin (an anthraquinone dye) is said to act irreversibly on this enzyme, binding covalently to it. This compound is also active on solid malignant tumors (Heath and Rock 2004).

4.2.1.9. *Isoleucine tRNA synthetase (EC 6)*

This is an enzyme capable of placing the amino acid isoleucine on the corresponding tRNA in the 3' position of the nucleic acid. The amino acid is first recognized by an enzyme domain, then activated by ATP. The result is an isoleucyl-AMP. The enzyme can then transfer the isoleucine to the tRNA, releasing the AMP. This recognition is made possible by a specific domain which locates the isoleucine anticodon of the tRNA.

Icofungipen acts on isoleucine tRNA synthetases in fungi and bacteria (Hasenoehrl et al. 2006).

4.2.2. *Endogenous targets*

4.2.2.1. *Alpha-amylase (EC 3)*

In humans, the carbohydrates contained in food are broken down during digestion into smaller molecules, such as dextrans or disaccharides (maltose, sucrose, lactose), for future absorption in the jejunum. This transformation involves two key enzymes in the digestive tract: salivary amylase produced by the salivary glands and pancreatic amylase acting in the duodenum.

Research into pancreatic pathologies are in progress, particularly in the field of enzyme substitution.

4.2.2.2. *Enzymes involved in galactose metabolism*

Galactosemia is a genetic disease resulting from the deficiency of enzymes involved in galactose metabolism. This hexose is a constituent of the milk disaccharide lactose (β-D-galactopyranosyl-D-glucose). Lactose metabolism involves three enzymes whose responsible genes have been identified. They enable galactose to be converted into glucose, as shown in Figure 4.1.

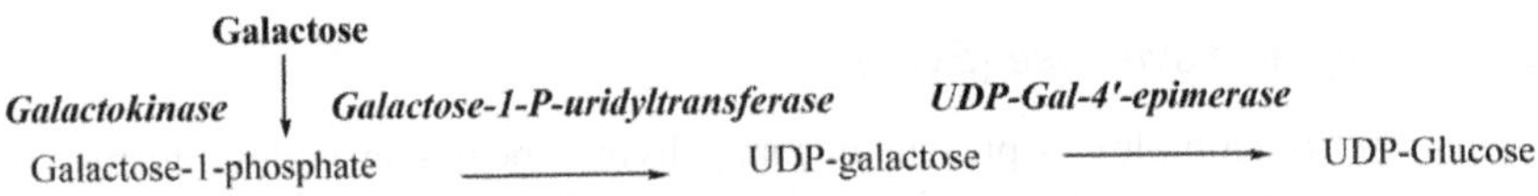

Figure 4.1. *Galactose metabolism*

The first stage transforms α-D-galactose into galactose-1-phosphate under the action of galactokinase (GALK), whose deficiency is responsible for galactosemia type 2, leading to cataracts. The second involves galactose-1-phosphate uridyltransferase (GALT), responsible for the majority of cases of galactosemia, known as classic galactosemia (type 1), manifested by liver, kidney or eye disorders that improve with a galactose-deprived diet, but cause neurological disorders in the long-term. Fatal if not treated with the appropriate measures, it mainly affects infants fed with milk that contains lactose. Treatment consists of avoiding galactose intake (lactose-deprived milk), administering antioxidants to reduce oxidative stress and developing treatments with substrate-replacing and substrate-reducing enzymes. The third stage involves galactose 4'-epimerase (GALE), responsible for galactosemia type 3, which is quite rare and can manifest in mild or severe forms (Touati and Ogier de Baulny 2005; Timson 2016; Demirbas et al. 2018).

4.2.2.3. *Enzymes involved in fructose metabolism*

Fructosemia is the result of enzyme deficiencies in the metabolism of fructose, a constituent of sucrose α-D-glucopyranosyl-β-D-fructofuranoside. Cane sugar contains it, and fructose is found in abundance in fruit, honey and pastries. Fructose intolerance is mainly the consequence of a deficiency in aldolase B (Figure 4.2), encoded by the ALDOB gene located on chromosome 9; fructokinase deficiency is less severe (Cox 1994).

The formation of glyceraldehyde-3-phosphate, the same intermediate of glucose metabolic activity, enables fructose metabolism to continue in the same way.

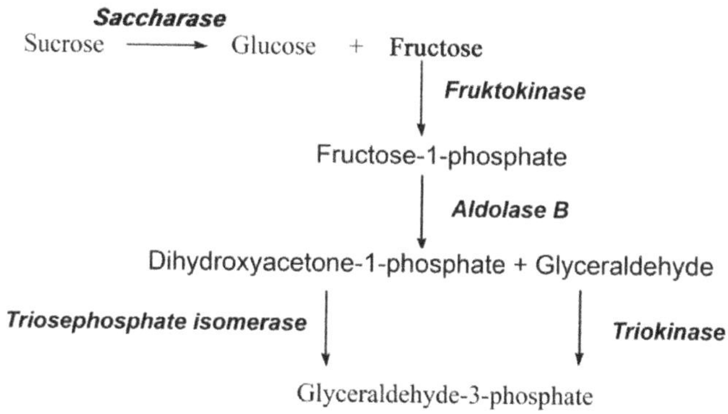

Figure 4.2. *Fructose metabolism*

4.2.2.4. *Arachidonate 5-lipoxygenase (EC 1)*

5-Lipoxygenase (5-LOX) is mainly expressed by leukocytes. It is responsible for the synthesis of leukotrienes. These lipid molecules are the main mediators of inflammation. 5-LOX can be inhibited by non-selective cyclooxygenase (COX) inhibitors (see Figure 3.17). Research into more specific inhibition of the enzyme is currently in progress.

4.2.2.5. *Farnesyl-diphosphate farnesyltransferase (EC 2)*

This enzyme, also known as squalene synthase, condenses two farnesyl pyrophosphate groups (15-carbon group) in the presence of magnesium and NADPH. This reaction is involved in cholesterol synthesis. Squalene synthase was studied for the treatment of certain hyperlipidemias and atherosclerosis (Kourounakis et al. 2011). It should be noted that the accumulation of farnesyl pyrophosphate resulting from this inhibition is toxic in animals. This family of cholesterol synthesis inhibitors needs to be approached with caution. This may explain the limited number of such compounds in clinical development, despite their appeal.

4.2.2.6. *Histone acetyltransferase (various isoenzymes) (EC 2)*

These enzymes are capable of transferring an acetyl group onto a histone protein. This acetylated protein, associated with DNA, modifies the compaction level of the genetic molecule. Acetylation promotes gene expression and cell proliferation, as well as inflammatory processes. In the context of gene expression deregulation, this is one category of enzymes that may be worth targeting. Inhibitors of histone acetylases and histone methyltransferases are already used.

4.2.2.7. *Phosphoribosylglycinamide formyltransferase (EC 2)*

This enzyme is involved in the purine synthesis pathway, specifically the formation of inosine monophosphate (IMP) from phosphoribosyl pyrophosphate. This catalyst is also known as glycinamide ribonucleotide formyltransferase (GARFT). This synthesis pathway involves 10 enzymes. Phosphoribosylglycinamide formyltransferase is the third enzyme in the process. It transforms glycinamide ribonucleotide into formylglycinamide ribonucleotide in the presence of tetrahydrofolic acid.

It should be noted that various purine synthesis pathways are known. These involve several enzymatic catalysts, each of which may serve as a potential therapeutic target. Pemetrexed acts as an antimetabolite, a multi-target antifolate that also acts on GARFT. It is used in the treatment of lung cancers, a point already addressed in relation to dihydrofolate reductase and thymidylate synthase, enzymes targeted in therapeutics.

Research is being carried out on GARFT in cancers (solid tumors). Unsuccessful clinical trials have been carried out with an inhibitor: Pelitrexol.

4.2.2.8. *Thyroxine-5'-deiodinase (EC 1)*

This enzyme is responsible for transforming T4 (thyroxine) into T3 (triiodothyronine). These two hormones play a major role in the body's development. T4 and T3 are thyroid hormones that bind to nuclear receptors. T3 is more active than T4. In the event of T3 overproduction, it is worth developing strategies to inhibit this enzyme, the last in the metabolic pathway.

NOTE.– Thyroperoxidase (EC 1), also involved in the T3 and T4 hormone anabolism pathway, can be targeted in the event of overproduction. It is the first enzyme in this pathway.

4.2.2.9. *Carboxypeptidases (several isoforms for each type of carboxypeptidase) (EC 3)*

These exopeptidases may belong to the metalloprotease, serine or cysteine protease classes. Each enzyme in this family has its own characteristics. For example, carboxypeptidases A are metalloproteases that specifically hydrolyze the C-terminal residue of protein when the radical is aromatic or hydrophobic.

These enzymes are involved in numerous biological processes, the main one being digestion. Their dysfunction is implicated in a variety of pathologies, including obesity, infertility, depression, epilepsy and neurodegeneration.

Despite research into carboxypeptidase A inhibitors, no therapeutic applications have come from it yet (Wentland et al. 2004).

4.2.2.10. *Dioxygenases (EC 1)*

Iron-containing enzymes such as heme enzymes play a significant role in biological systems, opening up new therapeutic perspectives. Three distinct heme-containing dioxygenases, tryptophan 2,3-dioxygenase (TDO), indoleamine 2,3-dioxygenase 1 (IDO1) and indoleamine 2,3-dioxygenase 2 (IDO2), catalyze the initial and limiting step of tryptophan catabolism via the kynurenine pathway in mammals. Overexpression of these enzymes leads to tryptophan depletion and accumulation of metabolic products, contributing to tumor immune tolerance and immune dysregulation in various pathologies.

Over the last few decades, a number of IDO1 inhibitors with high immunotherapy potential have been synthesized and assessed. Indoximod (2R)-2-amino-3-(1-methylindol-3-yl), propanoic acid (1-methyl-D-tryptophan) and an IDO1 peptide vaccine have reached the stage of clinical trials in certain treatments: glioma, melanoma, gliosarcoma and lung cancers (Yan et al. 2017).

4.2.2.11. *Caspases (EC 3)*

Caspases are hydrolases involved in numerous biological processes, notably as end-of-pathway actors in signal cascades, leading to apoptosis or in inflammation-related reactions.

The term “caspase” stands for Cysteinyl-dependent ASpartate ProteASE.

At least 15 caspase-type isoenzymes have been counted in humans (Shalini et al. 2015).

Research to discover selective inhibitors of each isoenzyme is underway, with little success to date. Potent inhibitors that have not led to therapeutic use include VX765, peptidomimetics and peptides (Zhang and Zheng 2016; Lee et al. 2017).

4.3. Rare diseases and enzymes

This therapeutic approach is related to certain enzymes involved in rare or orphan diseases. Among the latter involving enzymes, lysosomal pathologies provide an interesting framework for study[1].

An estimated 7,000 rare diseases are known today, 80% of which are originally genetic. Among these, lysosomal storage diseases account for over 50. They are caused by mutations in genes encoding abnormal lysosomal enzymes (National Institute of Health 2013; Orphanet 2013). The deficiency of a particular enzyme will lead to the accumulation of the corresponding substrate and give rise to the various phenotypic, physiological and biological manifestations of the disease (Ortolano et al. 2014; Pryde and Palmer 2014).

In addition, there are many difficulties involved, both in discovering the true cause of a rare disease and in implementing an appropriate treatment. Research must first identify the mutated or defective gene at the root of the disease. Comparisons of the genetic profiles of healthy subjects with those of a small number of patients are essential. The identification and separation of the gene(s) responsible, as well as the determination of the sequence of DNA and messenger RNA of the corresponding enzymes, is necessary to select a specific target, with a view to developing an appropriate therapy.

Research into rare diseases by pharmaceutical companies is often limited for economic and ethical reasons, not to mention the difficulties encountered in conducting clinical trials, given the limited number of patients involved. However, some rare drugs, known for this reason as “orphan drugs”, have been developed for certain diseases. They are often aimed at correcting the symptoms of the disease, and do not lead to a cure. However, knowledge of the human genome is opening up new prospects for gene therapies based on gene-editing technology. This technology uses modified endonucleases, specifically targeting the DNA sequence of the patient’s deficient or mutated gene, so as to enable it to be corrected. This therapeutic option is facilitated by the discovery of three types of endonucleases: zinc finger nucleases, TALENs (transcription activator-like effector nucleases) and CRISPR-Cas9 (clustered, regularly interspaced, short palindromic

1 For more information on these rare diseases in their entirety, please visit the following website: www.orpha.net.

repeat-associated-nucleases Cas9) (Yin et al. 2017). Clinical trials involving DNA transfer are progressively being developed, without any conclusive results to date (Shim et al. 2017).

In this area, the FDA has granted marketing authorization for Voretigene Neparvovec.rzyl.

4.3.1. *Lysosomal storage diseases*

Today, two main strategies are potentially used to treat patients with lysosomal enzyme deficiencies: enzyme replacement therapy (Brady 2006; Rohrbach and Clarke 2007) or substrate reduction therapy (Platt and Jeyakumar 2008). The former involves administering a substitute enzyme to replace the patient's defective enzyme and promote catabolism of the disease-causing substrate. However, its implementation presents a number of difficulties:

– Enzymes cannot generally be administered orally, as they are degraded by digestive enzymes.

– Intravenous administration is of limited effectiveness, due to the enzyme's difficulty in penetrating cell lysosomes.

– The enzyme used is less selective and less effective than the lysosomal enzyme, giving rise to side effects.

– Enzymes used to treat degenerative diseases of the central nervous system have difficulty crossing the blood–brain barrier, limiting their efficacy.

– There is a possibility of allergic reactions.

The second therapeutic approach involves the development of derivatives designed to reduce the storage substrate by inhibiting the enzyme involved in its biosynthesis. Examples include *Miglustat* and *Eliglustat* (see below), two synthetic molecules designed to inhibit glucosyl transferase and prevent glucocerebroside biosynthesis. Comprehensive information on the structure and metabolism of glycosphingolipids (GSLs), glycolipids and glycosaminoglycans (GAGs) involved in rare diseases can be found in publications dedicated to them (Sandhoff and Kolter 2003; Xu et al. 2010). Websites dedicated to these diseases can also be accessed. By way of example, we report a few cases of rare diseases resulting from genetic mutations, leading to deficiencies in certain enzymatic functions involved in the catabolism of GSLs, GAGs and other substrates. Orphan drugs that may have been developed for some of these pathologies involving enzyme deficiencies will be indicated. Promising gene therapy and stem cell approaches will not be detailed. Various pathologies are the consequence of abnormalities in the metabolism of lysosomal GSLs.

Figure 4.3 shows the various stages in their metabolism, leading to the accumulation of the corresponding substrate in the event of a deficiency in a particular enzyme.

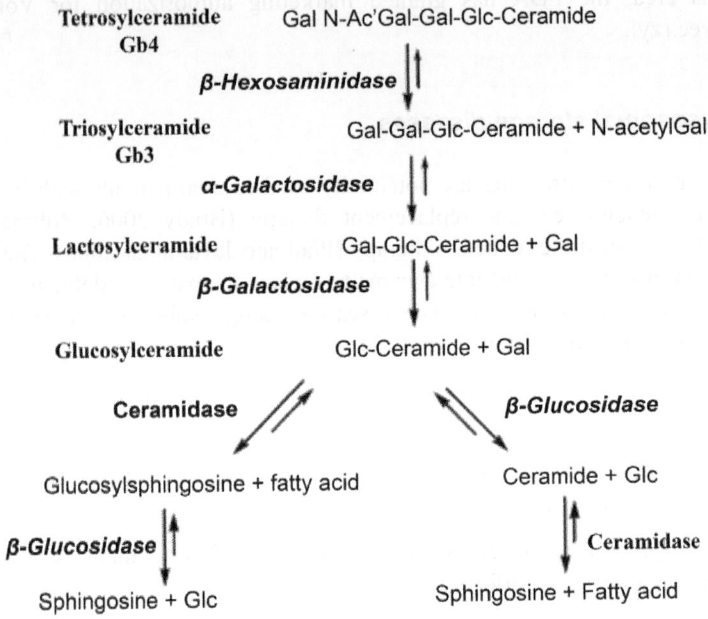

Figure 4.3. *Glycosphingolipid metabolism (catabolism enzymes). Gal: galactose; Glc: glucose; Gb: globoside*

Sandhoff disease is characterized by a defect in the action of β-hexosaminidase A/B, enabling the passage of Gb4 to Gb3, with the accumulation of Gb4, as well as (Ga-2 (asialoGM-2)) and GM-2, due to a deficiency in the same enzyme. The name of this pathology is gangliosidosis.

Tay–Sachs disease is the consequence of deficiency of the alpha sub-unit of hexosaminidase-A (Cachón-González et al. 2018).

The accumulation of GM-1 and GA-1, as well as keratan sulfate, is at the root of *Morquio syndrome*, characterized by a deficiency of β-galactosidase, also known as GM-1 gangliosidosis, for which there is currently no treatment.

In the case of *Fabry disease*, the defect in the transition from Gb3 to lactosylceramide (βGal-βGlc-Cer) (Figure 4.3) is impaired as a result of α-galactosidase A deficiency. This rare genetic lysosomal disease mainly results in the accumulation of globotriosylceramide (Gb3) in the lysosomes of many cell types. The disease manifests as renal and cardiovascular damage, as well as peripheral neuropathy (Spada et al. 2006). Treatment consists of promoting glycosphingolipid degradation through infusions of the following enzymes: *Agalsidase alfa* and *Agalsidase beta*.

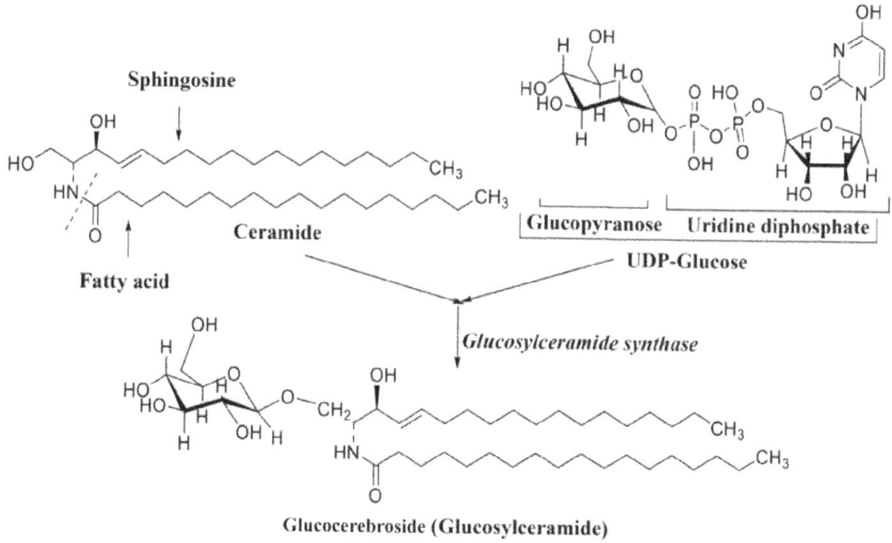

Figure 4.4. *Biosynthesis of glucocerebroside*

Gaucher disease, of which three types have been described, is an autosomal recessive hereditary disorder. It is characterized by a deficiency or absence of beta-glucocerebrosidase (β-glucosidase), which is responsible for the degradation of glucocerebroside (glucosylceramide) into ceramide (Figure 4.3). The disease results in lysosomal accumulation of glucosylceramide in macrophages of the liver, spleen, bone marrow and lymph nodes, as well as in red and white blood cell abnormalities and bone damage (Beutler 2006).

Figure 4.4 shows the biosynthesis of glucocerebroside (glucosylceramide) in the Golgi apparatus, by binding ceramide to β-D-glucose, supplied by uridine diphosphoglucose (UDP-glucose) under the action of glucosylceramide synthase.

Glucocerebroside (glycosylceramide) can also be formed from globosides or gangliosides by successive enzymatic degradation. Treatment of an accumulation of the molecule will have two objectives:

– inhibiting glucocerebroside biosynthesis by inhibiting glucosylceramide synthase (glucosyl transferase);

– degrading glucocerebrosides by restoring the activity of β-glucocerebrosidase (β-glucosidase acid), enabling ceramide formation, as shown in Figure 4.5.

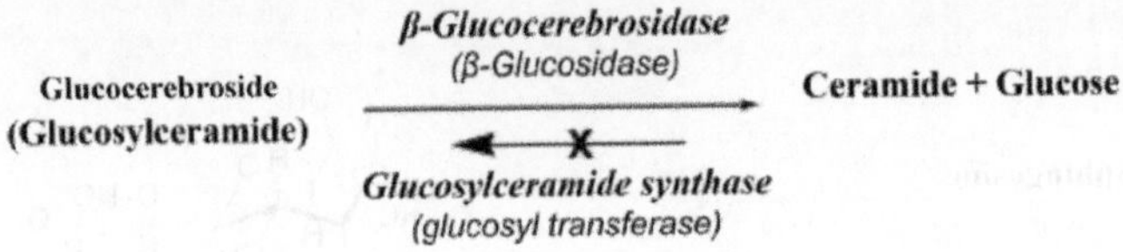

Figure 4.5. *Inhibition of glucocerebroside biosynthesis and activation of their degradation*

Enzyme replacement therapy uses *Imiglucerase* and *Alglucerase*, substitute enzymes for the hydrolysis of glucocerebrosides, in order to reduce their excess. Inactive orally (degradation), they are administered by intravenous infusion. The other approach, aimed at reducing the substrate, uses *Miglustat.* Its structure corresponds to a substituted piperidine triol, an N-alkylated structural analogue of D-glucose, corresponding to N-butyldeoxynorgyromycin.

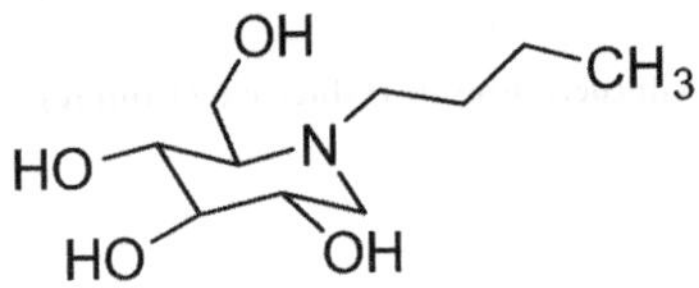

This compound is a reversible, competitive inhibitor of a glucosyl transferase, glucosylceramide synthase, responsible for the first step in the biosynthesis of most GSLs, whose levels are thus reduced. Unlike enzymes, it is administered orally. Similarly, *Eliglustat*, a structural analogue of glucosylceramide, inhibits glucoceramide synthase, which is involved in the biosynthesis of glucocerebrosides.

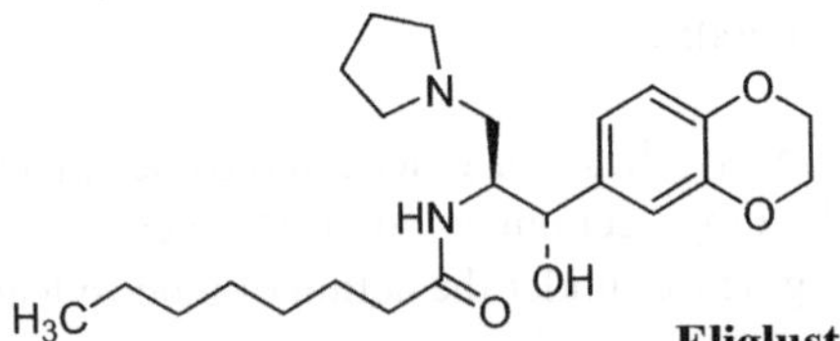

Eliglustat

NOTE.– Galactocerebroside biosynthesis takes place in the endoplasmic reticulum. A bond is created between a ceramide and the galactose supplied by uridine diphosphogalactose (UDP-galactose) through the action of ceramide galactosyltransferase. This reaction proceeds in the same way as shown in Figure 4.4 for glucocerebroside (Ichikawa et al. 1996). Galactocerebrosidase deficiency prevents the catabolism of galactocerebroside into ceramide and galactose, leading to the central nervous system disorders of *Krabbe disease*, for which there is currently no proper treatment.

Another genetic disease is *Farber's disease*. It results from the accumulation of ceramides in lysosomes due to a ceramidase deficiency, leading to an accumulation of lipids in joints and various tissues, including the central nervous system (Figure 4.6). There is currently no drug available to treat this disease.

Deficiency of a specific enzyme involved in GAG catabolism results in the accumulation of the corresponding undegraded substrate, leading to conditions known as mucopolysaccharidoses (MPS). Some diseases are also named after the person who first discovered them.

Dermatan sulfate and heparin sulfate enzyme deficiencies give rise to various pathologies, with the accumulation of the undegraded substrate corresponding to the enzyme concerned.

For example, deficiency of iduronidate sulfatase, an enzyme required for the first step in the catabolism of dermatan sulfate and heparan sulfate, is responsible for *Hunter syndrome*, corresponding to MPS II. One treatment for this disease involves administration of the enzyme produced by recombinant DNA technology known as *Idursulfatase* (Whiteman and Kimura 2017).

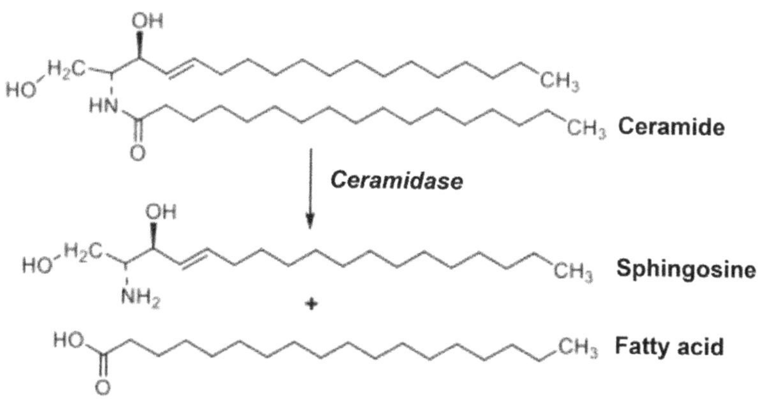

Figure 4.6. *Ceramide catabolism by ceramidase, the deficient enzyme in Farber's disease*

The second stage in the metabolism of these two GAGs involves iduronidase. Deficiency of this enzyme gives rise to the *Hurler–Scheie syndrome* or MPS I. Treatment of this disease involves the substitution enzyme *Laronidase.*

The third stage of metabolism involves N-acetyl-galactosamine-4-sulfatase in the case of dermatan sulfate, whose deficiency leads to the *Maroteaux–Lamay syndrome* or *MPS VI*. Treatment by enzyme substitution is carried out with *Galsulfase.*

In the case of heparin sulfate, the third stage of catabolism involves various enzymes. N-acetyl-galactosamine-6-sulfatase deficiency results in *Sanfilippo syndrome type D, MPS III*. Heparin N-sulfatase deficiency results in *Sanfilippo syndrome type A, MPS III*. Alpha-glucosaminide N-acetyl transferase deficiency leads to *Sanfilippo disease type C, MPS III*. The final stage in heparin sulfate catabolism involves alpha-N-acetyl glucosaminidase, the deficiency of which is responsible for *Sanfilippo disease type B, MPS III*. There is currently no suitable treatment for these diseases.

Chondroitin sulfates consist of disaccharide residues linked by a β1-3 bond and composed of a glucuronic acid (or its 2-O-sulfate derivative) and an N-acetylgalactosamine (or one of its 4-O-sulfate, 6-O-sulfate or 4,6-di-O-sulfate derivatives).

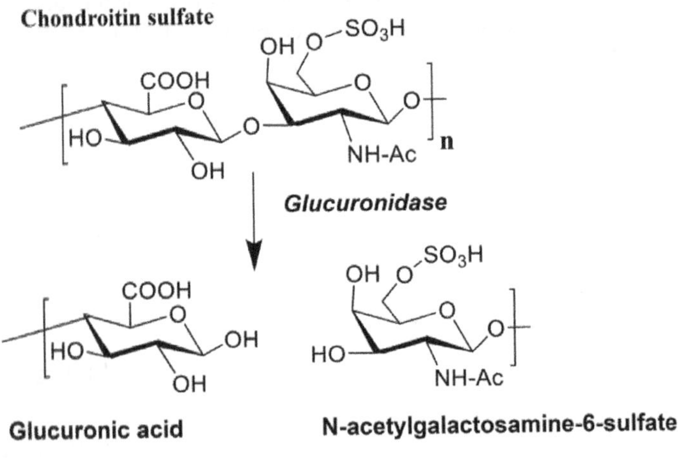

Figure 4.7. *Chondroitin sulfate metabolism*

The metabolism of chondroitin sulfate and keratan sulfate both involves N-acetyl-galactosamine-6-sulfatase (the first stage in the metabolism of these two substrates). N-acetyl-galactosamine-6-sulfatase (GALNS) deficiency during the first step of metabolism is the cause of *Morquio A syndrome*, mucopolysaccharidosis (MPS, IVA), an autosomal recessive disease (Tomatsu et al. 2014). It gives rise to an accumulation of glycosaminoglycans (GAGs): keratan sulfate (KS) and chondroitin-6-sulfate (C6S), mainly in cartilage and its extracellular matrix. The accumulation of undegraded and stored substrates has a direct impact on cartilage and bone development, resulting in skeletal dysplasia (Figures 4.7 and 4.8).

The subsequent steps in the metabolism of these two GAGs are different. Thus, in the case of chondroitin sulfate, beta-hexose amidinase deficiency during the second metabolic step gives rise to *Tay–Sachs disease*, and, subsequently, beta-glucuronidase deficiency results in *Sly syndrome, MPS VII*. There is currently no suitable treatment for this disease.

Keratan sulfate chains are made up of disaccharide residues linked by beta 1-3 and beta 1-4 bonds. They are composed of galactose (replacing hexuronic acid) and N-acetylglucosamine. 6-O-sulfate residues may be present on both oses, and in some cases, a fucose may be present in the C3 position of N-acetylglucosamine. Deficiency of beta-galactosidase during the second stage of keratan sulfate metabolism leads to *Morquio B disease, MPS IV-B*, for which there is currently no suitable treatment.

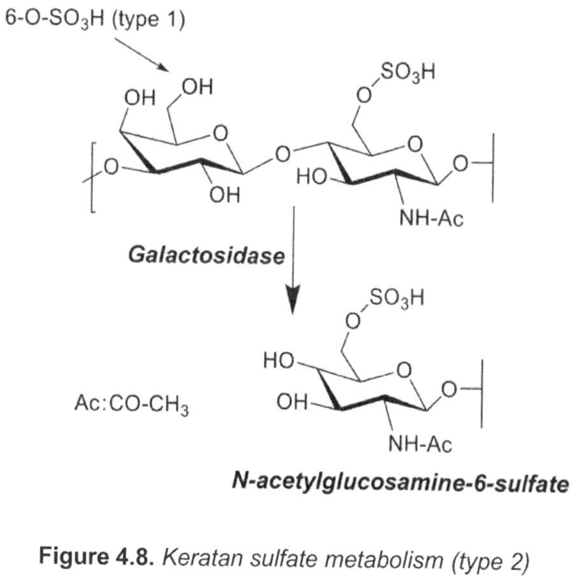

Figure 4.8. *Keratan sulfate metabolism (type 2)*

Hyaluronic acid is the only GAG without sulfate residues. It is made up of disaccharides linked by β1-4 bonds. Disaccharides are made up of glucuronic acid and N-acetylglucosamine, linked by a β1-3 bond.

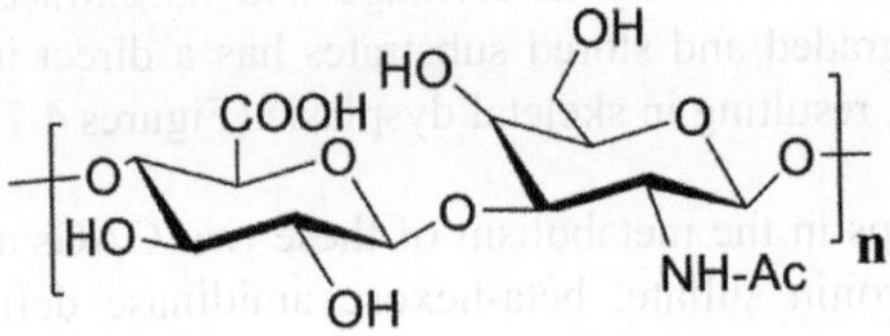

Hyaluronic acid

Hyaluronic acid is a constituent of cartilage tissue. It is used in intra-articular injections to increase synovial fluid, and in combination with other substances, such as glucosamine and its salts, in the treatment of osteoarthritis (Kirkiacharian 2010). These treatments are rarely used today due to new therapeutic options (interleukin and TNF-alpha (Tumor Necrosis factor) inhibitors).

Other rare pathologies include *Niemann–Pick disease*, of which there are essentially three variants: A, B and C. It results from genetic abnormalities that are transmitted in the autosomal recessive mode. It is characterized by a deficiency of sphingomyelinase in cell lysosomes. Under these conditions, sphingomyelin cannot be degraded, and accumulates in the cells, leading to cell death. The disease manifests itself through the malfunctioning of various organs (liver, spleen), as well as the respiratory and cardiovascular systems (types A and B). In type C of this disease, impaired lipid transport leads to cholesterol accumulation in the liver and spleen, as well as sphingolipid accumulation in neurons with neurological disorders (Vanier 2013; Lyseng-Williamson 2014). Treatment involves *Miglustat* with no subsequent cure.

Pompe disease is a genetic disorder caused by an abnormality in the metabolism of glycogen, a glucose polymer with α1-4 and α1-6 glycosidic bonds. It is one of a group of disorders known as glycogenoses (Vianey-Saban and Pirau 2010). It results from a deficiency in acidic lysosomal alpha-glucosidase (acid maltase), which enables the release of glucose (Glc) from glycogen and the formation of glucose-1-phosphate under the action of glycogen phosphorylase kinase. Under the action of phosphoglucomutase (phosphohexose mutase), glucose-1-phosphate then forms glucose-6-phosphate, an essential energy source for cellular metabolism. Subsequent enzymatic steps lead to glyceraldehyde-3-phosphate, and then pyruvate (Figure 4.9). The disease manifests itself in altered muscular and cardiac function.

Treatment consists of *Alpha-glucosidase* infusions, enabling glucose formation. Recently, enzyme replacement therapy with avalglucosidase alfa for long-term treatment of late-onset and infantile forms of the disease has been launched.

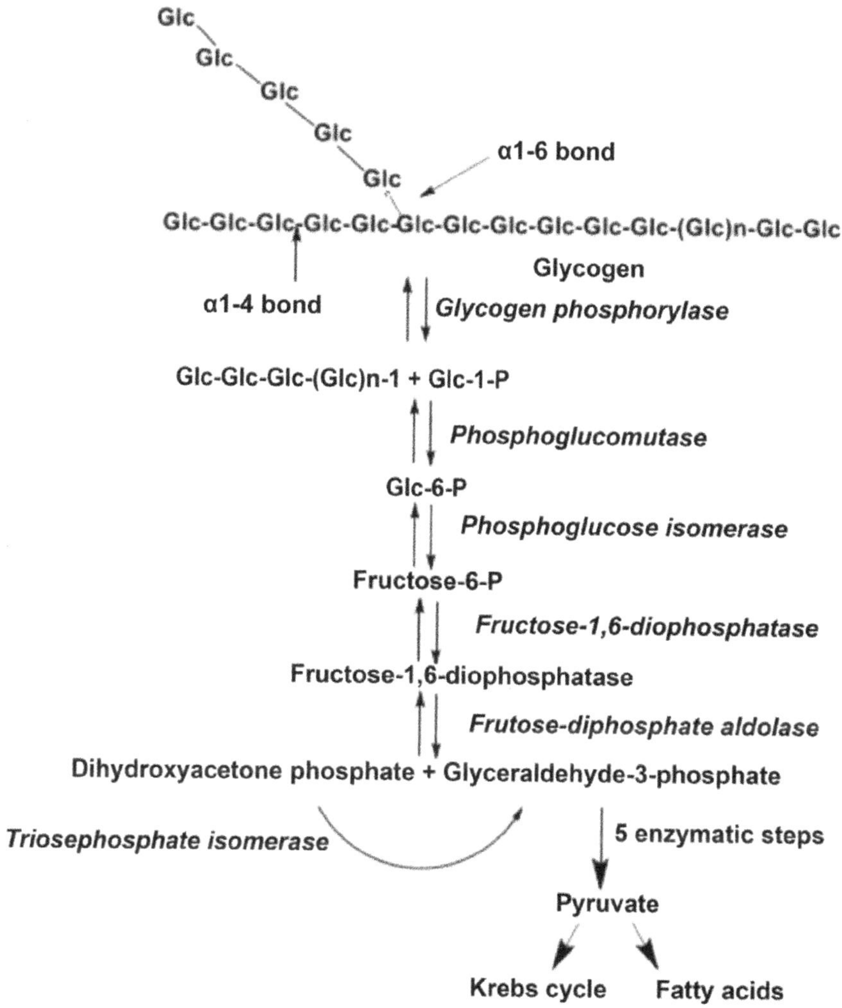

Figure 4.9. *Glycogen metabolism*

4.3.2. *Other rare genetic diseases linked to one or more enzymes*

Extremely rare enzyme-related genetic diseases include the following:

– *Hereditary orotic aciduria*. This is an autosomal recessive disorder. It is characterized by stunted growth, anemia and increased urinary excretion of orotic acid. The disorder is linked to a severe deficiency of uridine 5'-mono-phosphate (UMP) synthase. This catalyst of the pyrimidine biosynthesis pathway has two functions: orotate phosphoribosyltransferase activity (EC 2) and orotidine 5'-monophosphate decarboxylase activity (EC 4). This enzyme is encoded by a gene located on chromosome 3. No treatment currently exists.

– *Congenital adrenal hyperplasia* (CAH). This is a rare genetic disease linked to a variable loss of activity of certain adrenal enzymes. The medullary adrenals produce epinephrine, while the adrenal cortices produce cortisol and aldosterone, as well as androgens produced mainly by the testes. Most cases involve 21-hydroxylase deficiency. Deficiencies in 11-beta hydroxylase, 3-beta hydroxysteroid dehydrogenase or 17-alpha hydroxylase are possible, but rarer. In all cases, the disease is autosomal recessive. Hormone replacement therapy consists of providing patients with the deficient hormones. The production of excess hormones must also be curbed. Treatment must be carried out throughout life.

– *Primary hyperoxaluria* (PH). This rare disease may be linked to mutations in two genes. It affects one in every 120,000 births. The most common form (PH 1) is caused by a deficiency of the liver enzyme, alanine glyoxylate aminotransferase (AGT). This prevents the transformation of glyoxylate into glycine, leading to the overproduction of oxalate and the formation of kidney stones (calcium oxalate). Type 2 disease (PH 2) is linked to hydroxypyruvate reductase deficiency. This disease causes kidney lithiasis, resulting in regular abdominal pain, hematuria and urinary tract infections in affected children. In 50% of cases, the disease appears before the age of 5. Currently, the only treatment to lower oxalate levels is vitamin B6. When the mutated enzyme is present, this cofactor can increase its activity, thereby lowering oxalate levels. Other treatments aimed at solubilizing calcium oxalate are possible (sodium or potassium citrate or sodium bicarbonate to alkalize urine).

4.4. References

Beutler, E.C. (2006). Gaucher disease: Multiple lessons from a single gene disorder. *Acta Paediatr.*, 95, 103–109.

Brady, R.O. (2006). Enzyme replacement for lysosomal diseases. *Annu. Rev. Med.*, 57, 283.

Cachón-González, M.B., Zaccariotto, E., Cox, T.M. (2018). Genetics and therapies for GM2 gangliosidosis. *Curr. Gene Ther.*, 18(2), 68–89.

Camarasa, M.J., Velázquez, S., San-Félix, A., Pérez-Pérez, M.J., Gago, F. (2006). Dimerization inhibitors of HIV-1 reverse transcriptase, protease and integrase: A single mode of inhibition for the three HIV enzymes? *Antiviral Res.*, 71(2/3), 260–267.

Cox, T.M. (1994). Aldolase B and fructose intolerance. *FASEB*, 8, 62–71.

Demirbas, D., Coelho, A.I., Rubio-Gozalbo, M.E., Berry, G.T. (2018). Hereditary galactosemia. *Metabolism*, 83, 188–196.

Divita, G., Bayon, J.G., Ritinger, K., Cherman, J.C., Goody, R.S. (1995). Interface peptides as structure-based human immunodeficiency reverse transcriptase inhibitors. *J. Biol. Chem.*, 270, 28642–28646.

Dumond, J., Boggetto, N., Schramm, H.J., Schramm, W., Takahashi, M., Reboud-Ravaux, M. (2003). Thyroxine-derivatives of lipopeptides: Bifunctional dimerization inhibitors of human immunodeficiency virus-1 protease. *Biochem. Pharmacol.*, 65(7), 1097–1102.

Dumond, J., Tronchet, J.M., Kirkiacharian, S., Seman, M., Reboud-Ravaux, M. (2020). Insights into biophysical méthods to study interaction between HIV-1 reverse transcriptase and non-nucleoside reverse transcriptase inhibitors. *Let. Drug Des. Disco.*, 17(6), 818–825.

Franciskovich, J., Houseman, K., Mueller, R., Chmielewski, J.A. (1993). A systematic evaluation of the inhibition of HIV-1 protease by its C- and N- terminal peptides. *Bioorg. Med. Chem. Let.*, 3, 765–768.

Gao, Y., Yan, L., Huang, Y., Liu, F., Zhao, Y., Cao, L., Wang, T., Sun, Q., Ming, Z., Zhang, L. et al. (2020). Structure of the RNA-dependent RNA polymerase from Covid-19 virus. *Science*, 368(6492), 779–782.

Hasenoehrl, A., Gali, C., Ergovi, G., Mari, N., Mittendorf, N.J., Geschke, U., Schmidt, A., Schoenfeld, V. (2006). Vitro activity and in vivo efficacy of icofungipen (PLD-118), a novel oral antifungal agent, against the pathogenic yeast candida albicans. *Antimicrob. Agents Chemother.*, 50(9), 3011–3018.

Heath, R.J. and Rock, C.O. (2004). Fatty acid biosynthesis as a target for novel antibacterials. *Curr. Opin. Investig. Drugs*, 5(2), 146–153.

Ichikawa, S., Sakiyama, H., Suzuki, G., Hidari, K.I., Hirabayashi, Y. (1996). Expression cloning of a cDNAfor human ceramide glucosyltransferase that catalyzes the first glycosylation step of glycosphingolipid synthesis. *Proc. Natl. Acad. Sci. USA*, 93, 4638–4643.

Kirkiacharian, S. (2010). *Guide de chimie médicinale et médicaments*. Lavoisier, Paris.

Kourounakis, A.P., Katselou, M.G., Matralis, A.N., Ladopoulou, E.M., Bavavea, E. (2011). Squalene synthase inhibitors: An update on the search for new antihyperlipidemic and antiatherosclerotic agents. *Curr. Med. Chem.*, 18(29), 4418–4439.

Lee, H., Lee, J.H., Deoksoo, A., Kim, C.G., Kim, J.H., Kim, S. (2017). Caspases inhibitors: A review of recently patented compounds (2013–2015). *Expert Opin Ther Pat.* 28(1), 47–59. doi: 10.1080/13543776.2017.1378426.

Liu, Y., Wang, K., Massoud, T.F., Paulmurugan, R. (2020). SARS-CoV-2 vaccine development: An overview and perspectives. *ACS Pharmacol. Transl. Sci.*, 3(5), 844–858.

Lyseng-Williamson, K.A. (2014). Miglustat: A review of its use in Niemann-Pick disease type C. *Drugs*, 74, 61–74.

National Institute of Health (2013). Office of rare disease research [Online]. Available at: http://raredisease.Inf.gov/AboutUs.aspix.

Orphanet (2013). Site des maladies rares et des médicaments orphelins [Online]. Available at: http://www.orphanet.net/consor/cgibin/EducationAboutrareDiseases.php.

Ortolano, S., Viéitez, J., Navarro, C., Carlos Spuch, C. (2014). Treatment of lysosomal storage diseases: Recent patents and future strategies. *Recent Patents on Endocrine, Metabolic & Immune Drug Discovery*, 8, 9–25.

Perugini, M.A., Griffin, M.D., Smith, B.J., Webb, L.E., Davis, A.J., Handman, E. (2005). Insight into the self-association of key enzymes from pathogenic species. *Eur. Biophys. J.*, 34(5), 469–476.

Platt, F.M. and Jeyakumar, M. (2008). Substrate reduction therapy. *Acta Pediatr.*, 97, 88.

Pomel, S., Mao, W., Ha-Duong, T., Cavé, C., Loiseau, P.M. (2019). GDP-mannose pyrophosphorylase: A biologically validated target for drug development against leishmaniasis. *Front. Cell Inf. Micro.*, 9(186).

Pryde, C.D. and Palmer, M.J. (2014). *Orphan Drugs and Rare Diseases*. The Royal Society of Chemistry, Cambridge.

Rohrbach, R. and Clarke, J.T. (2007). Treatment of lysosomal storage disorders: Progress with enzyme replacement therapy. *Drugs*, 67, 2697–2716.

Sandhoff, K. and Kolter, T. (2003). Biosynthesis and degradation of mammalian glycosphingolipids. *Philos. Trans. R. Soc. Lond. B. Biol. Sci.*, 358(1433), 847–861.

Schechter, I. and Berger, A. (1967). On the size of the active site in proteases. I. Papain. *Biochem. Biophys. Res. Commun.*, 27(2), 157–162.

Schramm, H.J., Breipohl, G., Hansen, J., Henke, S., Jaeger, E., Meichsner, C., Riess, G., Ruppert, D., Rucknagel, K.P., Schafer, W. (1992). Inhibition of HIV-1 protease by short peptides derived from the terminal segments of the protease. *Biochem. Biophys. Res. Commun.*, 184(2), 980–985.

Shalini, S., Dorstyn, L., Kumar, S. (2015). Old, new and emerging functions of caspases. *Cell Death Dif.*, 22(4), 526–539.

Shim, G., Kim, D., Park, G.T., Jin, H., Suh, S.K., Oh, Y.K. (2017). Therapeuticgene editing: Delivery and regulatory perspectives. *Acta Pharmacol. Sin.*, 38, 738–775.

Skovpen, Y.V., Conly, C.J., Sanders, D.A., Palmer, D.R. (2016). Biomimetic design results in a potent allosteric inhibitor of dihydrodipicolinate synthase from campylobacter jejuni. *J. Am. Chem. Soc.*, 138(6), 2014–2020.

Spada, M., Pagliardini, S., Yasuda, M., Tukel, T., Thiagarajan, G., Sakuraba, H., Ponzone, A., Desnick, R.J. (2006). High incidence of later-onset Fabry disease revealed by newborn screening. *Am. J. Hum. Genet.*, 79, 31–40.

Stevens, M., Pannecouque, C., De Clercq, E., Balzarini, J. (2003). Novel human immunodeficiency virus (HIV) inhibitors that have a dual mode of anti-HIV action. *Antimicrob. Agents Chemother.*, 47(10), 3109–3116.

Timson, D.J. (2016). The molecular basis of galactosemia. Past, present and future. *Gene*, 589, 133–141.

Tomatsu, S., Yasuda, E., Patel, P., Rhunke, K., Shimada, T., Mackenzie, W.G., Mason, R., Thacker, M.M., Theroux, M., Montano, A.M. et al. (2014). Morquio A syndrome: Diagnosis and current and future therapies. *Pediatr. Endocrinol. Rev.*, 12(1), 141–151.

Touati, G. and Ogier de Baulny, H. (2005). Anomalies héréditaires du métabolisme du galactose et du fructose. *EMC – Pédiatrie*, 2, 151–161.

Tronchet, J.M., Apparu-Turquois, C., Laroze, N., Seman, M. (2002). Selectivity of the reverse transcriptase inhibitors toward types 1 and 2 of human immunodeficiency virus (HIV). *Ann. Pharm. Fr.*, 60(4), 227–231.

Vanier, M.T. (2013). Niemann-Pick diseases. *Handb. Clin. Neurol.*, 113, 1717–1721.

Vianey-Saban, C. and Pirau, M. (2010). Les glycogénoses. *Revue Francophone des Laboratoires*, 425, 39–52.

Wentland, M.P., Raza, S., Gao, Y. (2004). 96-wellplate colorimetric assay for Ki determination of (+-)-2- benzylsuccinic acid, an inhibitor of carboxypeptidase A: A laboratory experiment in drug discovery. *J. Chem. Educ.*, 81, 398–400.

Whiteman, D.A. and Kimura, A. (2017). Development of idursulfase therapy for mucopolysaccharidosis type II (Hunter syndrome): The past, the present and the future. *Drug Des. Devel. Ther.*, 11, 2467–2480.

Xu, Y.-H., Barnes, S., Ying Sun, Y., Grabowski, G.A. (2010). Multi-system disorders of glycosphingolipid and ganglioside metabolism. *J. Lipid. Res.*, 51, 1643–1675.

Yan, D., Lin, Y.W., Tan, X. (2017). Heme-containing enzymes and inhibitors for tryptophan metabolism. *Metallomics*, 9(9), 1230–1240.

Yin, H., Kauffman, K.J., Anderson, D.G. (2017). Delivery technologies for genome editing. *Nat. Rev. Drug Discov.*, 16, 387–399.

Yu, F., Liu, X., Zhan, P., De Clercq, E. (2008). Recent advances in the research of HIV-1 RNase H inhibitors. *Mini-Reviews in Med. Chem.*, 8(12).

Zhang, Y. and Zheng, Y. (2016). Effects and mechanims of potent caspase 1 inhibitor VX765 treatment on collagen-induced arthritis in mice. *Clin. Exp. Rheu.*, 34(1), 111–118.

Zhou, Q.A., Kato-Weinstein, J., Li, Y., Deng, Y., Granet, R., Garner, L., Liu, C., Polshakov, D., Gessner, C., Watkins, S. (2020). Potential therapeutic agents and associated bioassay data for COVID-19 and related human coronavirus infections. *ACS Pharmacology & Translational Science*, 3(5), 813–834.

5

Which "New" Drugs are We Moving Towards Now?

The techniques or molecules detailed have direct applications in the pharmaceutical industry in general and particularly in enzymology.

5.1. Chemistry

At the heart of medicinal enzymology is the chemistry needed to generate the molecules of interest and distribute them throughout the body.

5.1.1. *Combinatorial chemistry (Duléry 2007)*

In addition to combinatorial chemical synthesis, parallel or high-throughput synthetic routes can also be mentioned. These chemical synthesis techniques are used to produce molecules that act on enzymes.

Combinatorial chemistry, like the other two methods, can be used to generate libraries of compounds that are essential to the discovery of new active ingredients.

The *Journal of Combinatorial Chemistry* is entirely dedicated to this field of chemistry.

Historically, classical chemical synthesis brings together two reactants, X and Y, in a single reaction to produce a product X-Y. This product must be obtained in the highest possible yield. In combinatorial synthesis, multiple reactions are carried out simultaneously. These reactions involve families of reagents. Typically, a set of compounds of reactivity X (X1...Xn) is combined with a set of compounds of reactivity Y (Y1...Yn) to obtain, in this case, n^2 molecules, at least in theory. If the compounds of reactivity X and Y are not equal, there are n compounds X and n' compounds Y, thus leading the number of combinations generated to n.n'. As these experiments already produce a large number of products, it is vital to avoid secondary reactions that could lead to numerous molecules being formed that would then be impossible to characterize. Therefore, a highly efficient reaction chemistry is needed.

Different compound selection strategies can be implemented.

One involves bifunctional reactive units. Reactions are successive, with each unit reacting with one of the functions of the next. Another strategy is to use multifunctional compounds. In this case, the Bn, Cn and Dn units react with a central derivative An (known as core or scaffold) featuring several different functions, each specific to the Bn, Cn and Dn units. Ultimately, the functionality of the units can be changed or modulated during the involved reactions. After the reaction of an An unit with a Bn unit, the reactivity of An can be modified, allowing Cn to react with An-Bn to form the product Bn-An-Cn. There is a reaction timeline. If Bn does not react with An, then Cn cannot interact.

This field of chemistry, which has already been widely exploited for three decades, is beginning to show its limitations. This calls for the development of other strategies to produce active molecules. In addition, given the very large number of generated compounds, it is necessary to test them on all possible targets and verify their specificity. This is not always easy when the primary objective is to create a molecular library, rather than to generate a molecule for a given target.

5.1.2. *Vectorization (CNRS 2017)*

The aim of vectorization is to generate various types of vectors (viral or synthetic) to transport active substances (drugs) within the body to target dysfunctional cells. The aim of this approach is to avoid affecting healthy cells that do not need the drug. It also makes it possible to control the release of the molecule in space and time. The field of vectorization combines chemistry, physics and biology.

Nanovectors can take on various forms:

– Nanospheres are spherical and biodegradable polymers. The active substance is integrated into the polymeric structure, forming a network. It is released when the structure is degraded.

– Nanocapsules are thin membranes of biodegradable polymers that contain the active substance. This is known as a "reservoir".

These nanovectors can have a specific surface marker, enabling them to accurately target abnormal cells.

Historically, there have been different generations of vectors:

– The first generation corresponds to assemblies of lipid monomers. These nanovectors include liposomes. They are spheres with a double phospholipid base, delimiting an aqueous compartment containing the active principle, if hydrophilic. If the molecule is hydrophobic, it will be embedded in the lipid membrane. This first generation of vectors is rapidly recognized as non-self by the immune system. The structure is taken over by Kupffer cells in the liver, where it is degraded and eliminated by phagocytosis. These vectors can therefore be used in liver pathologies (notably liver cancer), since the substance is indirectly targeted for delivery to this organ.

– The second generation corresponds to vectors based on a hydrophilic polymer, such as PolyEthyleneGlycol (PEG). When the vector contains PEG, it is referred to as a PEGylated vector. The drug is dispersed or dissolved in the polymer matrix. These structures are less immunogenic than first-generation ones. They can cross the liver barrier to reach other organs.

– The third generation of nanovectors features, in addition to the polymer, a molecular marker for targeting abnormal cells. This may be a ligand, a receptor or an antibody. These are known as PEGylated and decorated vectors. For example, the marker may be folic acid. These vectors can be used to target cancer cells displaying a specific protein membrane marker.

Nanovectors are currently used in numerous drug formulations. AmBisome® is a liposomal formulation of amphotericin B, an antifungal agent. Nanoparticles are used in the treatment of certain cancers, notably to improve the poor solubility of certain active ingredients. *Abraxane®* is a vector used in the treatment of breast cancer. It contains the active ingredient *Paclitaxel*. *Caelyx®* represents PEGylated liposomes of *Doxorubicin*. It is used to treat a wide range of cancers. In the field of enzymology, such vectors can be used to target cells that present catalyst dysfunction. It can also help overcome solubility problems encountered with certain types of molecule.

5.2. Biology

Biology makes it possible to identify new molecular targets or new classes of agents capable of acting on enzymatic catalysts, whether directly or indirectly.

5.2.1. *Inducers and repressors (Dion et al. 2007)*

Inducers and repressors are molecules known as transcription factors. They intervene at the level of gene transcription, promoting transcription in the case of inducers or limiting it in the case of repressors. Over 2,000 transcription factors have been identified in the human genome. Among them, a remarkable family of almost 1,000 members features characteristic zinc finger domains, motifs that interact with DNA. These proteins are involved in the transcriptional regulation of a large number of genes. There are many different types of zinc finger domains.

The Cys_2His_2 zinc finger domain is characteristic of transcription factors. It consists of approximately 30 amino acids with a consensus sequence Y-F-X-*C*-X2-5-*C*-X3-Y-F-X5-L-X2-*H*-X3-5-*H* (X is any amino acid). These amino acids have a secondary $\beta\beta\alpha$ structure, stabilized by a zinc ion. The divalent cation Zn^{2+} is coordinated by cysteines (in the β sheet) and histidines (in the α helix), which are conserved from one protein to another. This domain is notably found in the sub-family of nuclear receptors which, once a ligand is attached, dimerize to activate the transcription of specific genes. This sub-family comprises some 50 members. Their ligands have lipophilic structures. Ligands that are specific to all receptors in this sub-family have not been identified.

The use of these zinc finger proteins to correct the defective expression of certain genes (including enzyme-coding genes) is highly promising in the field of therapeutics.

In eukaryotes, it should be noted that other zinc finger domains are among the most common in protein families other than DNA-binding transcription factors.

Other DNA-binding protein domains include helix-turn-helix structures and leucine zippers.

5.2.2. *Antibodies (Broutin and Watier 2016)*

The pharmaceutical industry has been developing the use of monoclonal antibodies for treatment purposes for more than 30 years. As a result of their origin, therapeutic antibodies are referred to as biomedicines. They make up a growing proportion of the drugs approved for the market.

An antibody is a complex glycoprotein produced by the immune system, more specifically by activated B lymphocytes (plasma cells), in response to stimulation by a generally foreign component, also known as an antigen. These protein molecules are made up of two heavy and two light chains, forming a constant domain (Fc) and two reactive variable domains (Fab). The Fabs form paratopes that recognize a specific antigenic determinant (epitope). The four chains are covalently linked by disulfide bonds. Antibody paratopes bind to antigenic epitopes to neutralize them. Through their constant domain, antibodies are also able to recruit effectors such as macrophages, natural killer (NK) lymphocytes or the complement system, that are capable of eliminating pathogenic microorganisms.

Historically, the first monoclonal antibodies used as therapeutic agents in humans were murine molecules. As these foreign molecules caused significant humoral immune reactions (human antibodies acting against murine antibodies) and inactivation, humanization processes were developed. Starting with murine antibodies (-(m)omab®), science succeeded in obtaining chimeric human/mouse antibodies (-ximab®), then fully humanized mouse antibodies, except for the paratopes (-zumab®), and finally, fully humanized antibodies ((m)-umab®).

Interest in fully human antibodies lies in the fact that they act selectively on their target, increasing treatment efficacy and reducing side effects. Logically, immunology, hematology, oncology and rheumatology are the medical specialties that first developed and used these therapeutic antibodies. Today, they are the subject of research in almost every field of medicine. They are revolutionizing the medical sector, offering new therapeutic perspectives for previously incurable diseases. Today, there are over 50 therapeutic monoclonal antibodies with marketing authorization in oncology, rheumatology, hematology, dermatology, genetic diseases, pneumology, ophthalmology, neurology and many other fields.

The next few years will undoubtedly see the significant development of antibodies in a variety of formats, given that research is abundant and the number of targeted antigenic determinants associated with different antibody forms is practically infinite.

As far as antibodies are concerned, there is also a therapeutic technique that uses enzymes to release active ingredients under the direction of antibodies (which is of interest). Similar gene-directed or virus-directed techniques also exist:

– Antibody Directed Enzyme Prodrug Therapy (ADEPT), using precursors that release their active ingredient on the target under the action of an enzyme and are directed by antibodies.

– Gene-Directed Enzyme Prodrug Therapy (GDEPT), using precursors that release the active ingredient on the target under the action of an enzyme.

– Virus-Directed Enzyme Prodrug Therapy (VDEPT), using precursors that release the active ingredient on the target under the action of an enzyme and are directed by a virus.

In addition, the use of IgM to target cells to release a therapeutically targeted enzyme is being used to treat a variety of hematological, rheumatological and cancerous pathologies. This technique involves antibody–enzyme fusion (AEF). It has been applied to the treatment of Pompe disease, through the selective release of a glycogen-degrading enzyme into the cell's lysosome (Zhou et al. 2019).

5.3. Genetics

Genetics is a science that has the potential to correct the programming errors at the root of genetic pathologies. Dysfunctional enzyme genes are among the targets. However, we still have a long way to go to understand and control the global functioning of the genome and to ensure that a one-off genetic addition or replacement does not generate a cascade of harmful effects.

5.3.1. *Gene therapy (Inserm 2018)*

The first strategy involves importing a copy of a functional gene into a target cell, so that it can be expressed there and enable production of the protein that is lacking in a monogenic pathology. The "normal" gene is added to the host genome using a vector, without affecting the defective gene. This addition can be carried out directly on the patient, or on the patient's own stem cells, which are then reinjected into the patient. Working ex vivo enables us to control the steps more effectively, use limited quantities of vectors and avoid spread of the treatment to non-target organs. However, for many pathologies, the gene transfer protocol is carried out in vivo, by injecting the vectorized gene directly into the organism or organ to be treated, like a drug.

The second strategy involves eliminating or repairing an altered gene directly in the cell. This is called genome editing. This technique involves importing specific material into the cells "to be corrected", such as restriction enzymes (EC 3 nucleases, like ZFN (zinc finger nucleases)) and TALENs (transcriptional activator-like effector nucleases), in order to eliminate or specifically correct the defective gene. The "normal" gene must also be imported to correct the mutation causing the pathology. The CRISPR-Cas9 system, which is based on this approach, is very promising. Numerous clinical trials based on this system are already underway. A few paragraphs will be devoted to these specific nucleases later on.

The third strategy is developed for complex pathologies, where there is no single gene to repair or replace. In such cases, indirect strategies can be devised. By combining cell and gene therapies, cells with new therapeutic properties can be obtained.

The fourth strategy is to use genetically modified lytic viruses to kill targeted cells. These are essentially cancer cells.

After strategies, it is necessary to talk about vectors, given that they are involved in over 75% of gene therapy clinical trials. Viral vectors are the most common. These are modified viruses used to deliver a therapeutic gene to target cells. The viral vectors used are either integrative or non-integrative. Over the past 30 years, considerable engineering work has been carried out on these acellular viruses. Adeno-associated virus (AAV) and lentiviral vectors have largely replaced the earlier adenoviral and gamma-retroviral vectors. Simultaneously, the development of non-viral vectors and associated gene delivery techniques for genome editing is expanding. The aim of this research is to meet requirements for security and simplicity of use. For the time being, these approaches are less efficient than those using viral vectors.

Gene therapy therefore holds great promise in curing many human diseases. It brings together a number of biological processes that can be exploited to target or replace a defective gene. It is important to devote a few paragraphs to some invaluable tools: nuclease enzymes (EC 3), driven by guidance systems. These tools target a precise DNA sequence. They are among the most promising players in the adaptation of natural systems for genome-editing research. Rather than genomic editing, this would be correction, or genomic modification, as has already been mentioned.

The discovery of CRISPR sequences and the associated Cas9 enzyme at the adaptive immune response level in prokaryotes, as well as the use of this system in gene "editing", are revolutionizing molecular biology. Cas9 is a nuclease (EC 3) capable of hydrolyzing specific sequences (CRISPR) in DNA. Its use for research purposes is straightforward, allowing protocols to be devised with a high degree of flexibility in different fields, such as the development of specific cell lines, the mechanisms of occurrence of certain genetic pathologies or the identification of therapeutic targets.

Before returning to the CRISPR-Cas9 system, the description of nucleases that preceded it, zinc finger nucleases (ZFNs) and transcription activator-like effector nucleases (TALENs), is needed.

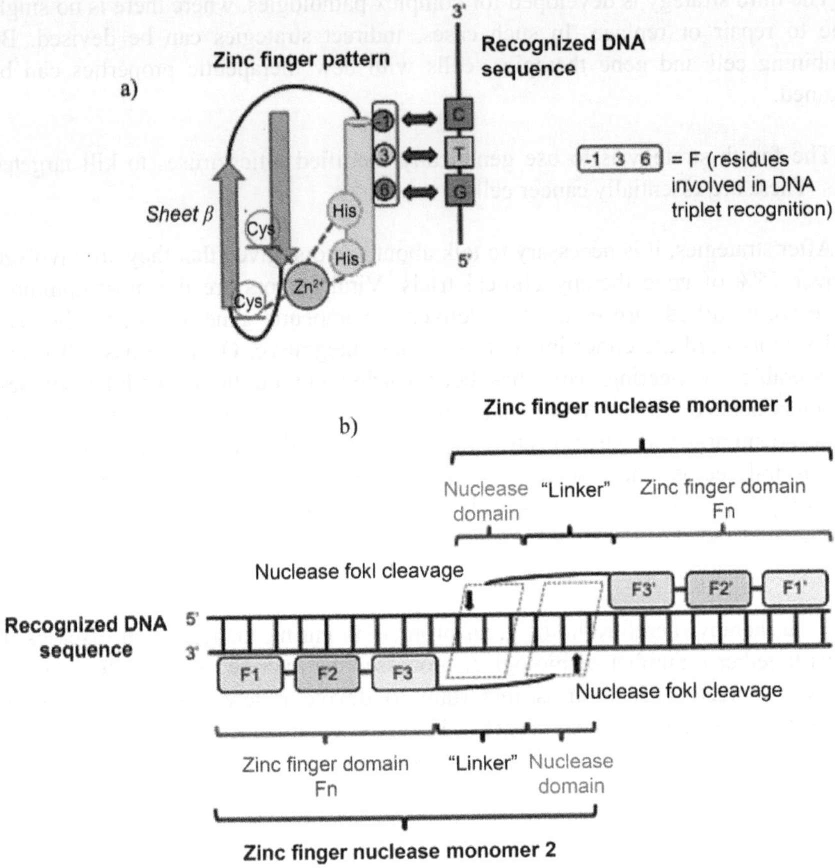

Figure 5.1. *a) Cys2-His2 zinc finger motif. Amino acid residues -1, 3 and 6 (Fn) respectively recognize the 3', central and 5' nitrogenous bases of the targeted DNA triplet. b) Positioning of the two zinc finger nuclease (ZFN) monomers, recognizing specific DNA sequences for cleavage at the restriction site. For a color version of this figure, see www.iste.co.uk/kirkiacharian/enzymes.zip*

In 2002, the first specific sequences recognized by ZFPs (zinc finger proteins) were reported. Based on these data, ZFNs were conceived. The latter are fusion proteins consisting of a Cys2-His2 zinc finger domain, characteristic of zinc finger proteins and a non-specific nuclease domain corresponding to a DNA-cleaving restriction enzyme. These two domains are linked by a "spacer" or "linker". The nuclease part is derived from a fokI (fok1) endonuclease. In eukaryotes, zinc finger domains are standard for the regulation of gene transcription or protein binding. In ZFNs, the challenge is to obtain zinc finger domains that are highly specific to DNA sequences, so that "genetic nanosurgery" can be performed with the greatest possible precision for the insert to be placed, the fragment to be removed or reversed, without disrupting the activity of the host genome. However, this is not an easy goal, as the panel of sites to be selected remains limited. Figure 5.1 describes the recognition of a specific DNA sequence by the zinc fingers, as well as the positioning of the ZFN dimer at the level of a DNA molecule to be cleaved. One monomer of the enzyme recognizes a strand of the molecule at a given level thanks to the specific zinc finger portion, perfectly positioning the nuclease domain to hydrolyze the molecule at the fokI restriction site. The second monomer is similarly positioned on the complementary strand, on the opposite side, so that its fokI domain is associated with the fokI domain of the other monomer for cleavage (Dion et al. 2007; Carroll 2011).

The study of TALEs began in 2005. These are natural DNA-binding proteins discovered in *Xanthomonas* sp. bacteria, containing 33–35 amino acid repeat motifs forming a DNA-binding domain. Shortly after the discovery of TALEs, research led to the synthesis of fusion proteins between TALEs and the fokI nuclease, like for ZFNs. TALENs were then born. The extra N indicates that they are nucleases. Unlike ZFNs, TALENs are easier and less time-consuming to create. In addition, these enzymes proved to be more specific and less cytotoxic than ZFNs. However, the high level of repetition in TALEN sequences and their genetic encoding, which is three times greater than that of ZFNs, presents problems for their administration and use. Figure 5.2 details the specificity of TALE sequences and the binding mode of a TALEN dimer, as well as the cleavage performed at the DNA sequence level (Dupret and Angrand 2014; Chandrasegaran and Carroll 2016).

The CRISPR-Cas9 system has been widely studied since the 2010s. It is currently the most sophisticated gene-editing system for correcting and modifying genetic sequences. In this system, unlike ZFN and TALEN, it is no longer protein sequences that target DNA, but a short RNA sequence. The use of this system avoids the need for lengthy protein engineering protocols to develop a specific restriction site for a targeted DNA sequence.

The history of CRISPR began in the 1990s, with the discovery of unusual repeated sequences that were separated from unrepeated variable sequences, the linkers (Ishino et al. 1987; Mojica et al. 1995). Partially palindromic repeat motifs were first described in *E. coli*. They were then analyzed in a large number of prokaryotes, with their role remaining undetermined. However, the name CRISPR was given to all these sequences. In the early 2000s, these sequences were linked to specific genetic sequences. Jansen established the CRISPR-associated system Cas, linking repeated and unrepeated sequences with genetic *loci*, forming CRISPR genetic sequences. These sequences are involved in DNA metabolism and gene expression. However, their precise functions were a mystery. In 2005, studies revealed that unrepeated sequences within CRISPRs are derived from bacteriophage DNA (linker sequences). Bacteria containing these sequences are known to resist infection by these phages. This last element shows that the CRISPR system is an adaptive immune response of prokaryotes to a virus. These sequences are an immune memory used by Cas proteins, which act as guides to scan foreign DNA sequences and deactivate them with their endonuclease function.

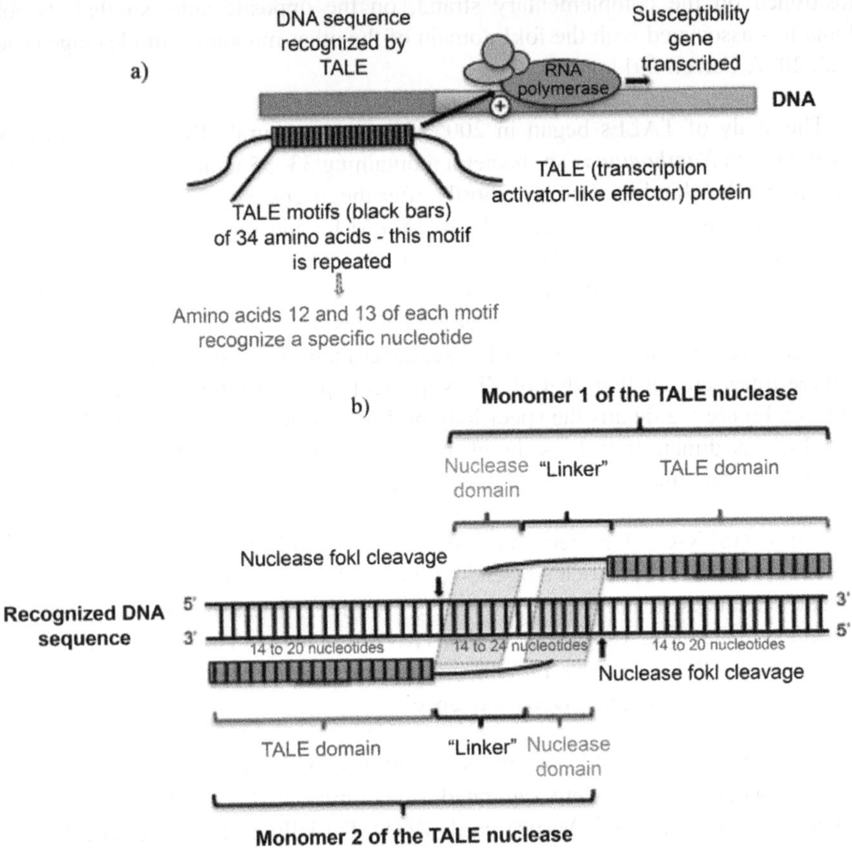

Figure 5.2. *a) TALE motif. Amino acid residues 12 and 13 of each motif recognize a nucleotide in the DNA sequence. TALE proteins are transcription factors produced by pathogenic bacteria of the genus* Xanthomonas. *b) Positioning of the two monomers of the TALE nuclease (TALEN) recognizing specific DNA sequences for cleavage at the restriction site. For a color version of this figure, see www.iste.co.uk/kirkiacharian/enzymes.zip*

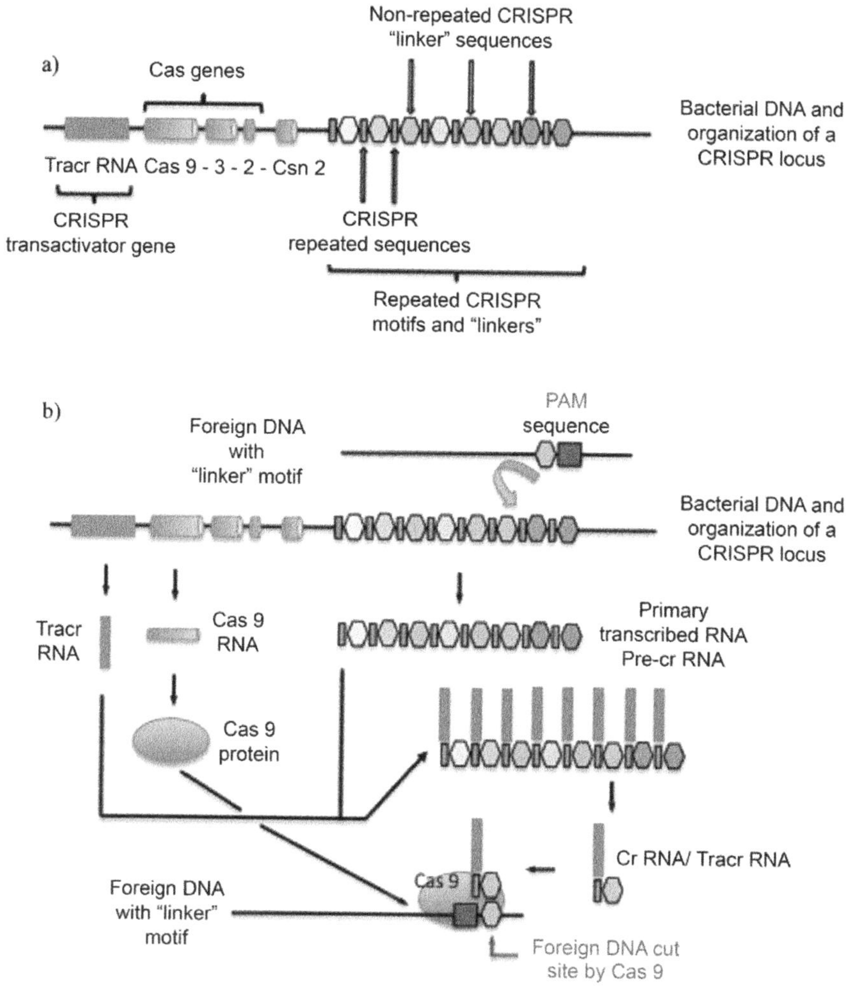

Figure 5.3. *How the CRISPR/Cas9 system works. For a color version of this figure, see www.iste.co.uk/kirkiacharian/enzymes.zip*

COMMENTS ON FIGURE 5.3.– *a) CRISPR-Cas9 system: tracr RNA gene (CRISPR transactivator), Cas9 gene and CRISPR motifs repeated with linker. b) Expression of the CRISPR system. Transcription yields tracr RNAs, the primary pre-cr transcribed RNA, then complexes (crRNA/tracr RNA) and Cas9 mRNA. The latter RNA is translated into the Cas9 protein. The Cas9 protein is positioned at the crRNA/tracrRNA level, recognizing the foreign DNA "linker" motif. The Cas9 nuclease can then cleave the recognized linker motif in the foreign DNA.*

At the genome level, the CRISPR *locus* is made up of three components:

– the CRISPR transactivator gene encoding a tracr RNA (CRISPR transactivator RNA);

– the Cas gene;

– CRISPR repeats and linkers.

All three are transcribed. The tracr RNA corresponds to a small RNA that matches each pre-crRNA repeat (pre-CRISPR RNA) to form a double-stranded RNA (tracr RNA/crRNA). The Cas protein has nuclease activity. Once the RNAs have been combined (the tracr RNAs/pre-crRNAs), an RNAse cleaves the pre-crRNA, corresponding to the CRISPR template, into several unitary tracr RNAs/pre-crRNAs, each of which contains a unique sequence of an invasive fragment. Like guide RNAs, tracr RNAs/crRNAs enable the endonuclease, which is carried by the Cas protein, to act in the right place to eliminate invasive viral or plasmid motifs. How the CRISPR/Cas9 system works is shown in Figure 5.3 (Jinek et al. 2012; Lino et al. 2018).

The use of the CRISPR-Cas9-based gene-editing technique has led to over 6,000 publications (the source used here is from 2018). Publications are still continuing today. In addition, over 3,000 genes are associated with mutations causing human pathologies. These two figures alone demonstrate the interest and usefulness of this system for correcting the errors at the root of genomic failures. The CRISPR/Cas9 gene-edited therapy was recently approved by EMA for the treatment of patients for severe sickle cell disease (denomination Exagamglogene autotemcel).

In conclusion, gene therapy has made it possible to develop two main types of treatment:

– molecules often embedded in vectors;

– innovative, more recent biological products, including antisense nucleotides, aptamers and small interfering RNAs. The next section will focus on antisense strategies at the messenger RNA level.

Among the drugs that have been brought to the market using these techniques, the following, non-exhaustive list can be mentioned:

– For gene therapy with vectors:

- *Gendicine*, marketed in China, is an adenovirus expressing the p53 gene. This drug is used to treat certain head and neck cancers.

- *Glybera* was launched to treat familial lipoprotein lipase deficiency. It is an adeno-associated vector. Due to a lack of demand, it was withdrawn from the market.

- *Luxturna* is an adeno-associated vector, indicated for retinal dystrophy linked to a specific RPE65 mutation. It was recently approved in the United States.

– Spinal muscular atrophy (SMA) treatment is now achieved by gene therapy using Onasemnogene abeparvovec ZOLGENSMA®. The treatment involves the administration of the gene SMN1 introduced in a modified virus (AAV9).

– For the combination of cell and gene therapies, Strimvelis is the very first drug in this field. It is indicated for severe combined immunodeficiency, linked to adenosine deaminase deficiency (Flinn and Gennery 2018);

– For lytic viruses targeting "diseased or malignant" cells, there is *Imlygic*, a lytic herpes virus indicated in the treatment of certain melanomas. *Oncorine*, marketed in China, is an oncolytic adenovirus used in the treatment of refractory nasopharyngeal cancers. It is used in combination with chemotherapy.

Many gene therapy clinical trials have been conducted since 1989. As with humanized antibodies, there is no doubt that advances in knowledge will enable the development of increasingly reliable and safe treatments over the coming decades, as long as all the recommendations stipulated by independent national and international health authorities are strictly adhered to.

5.3.2. *Antisense strategies (Hélène and Saison-Behmoaras 1994)*

Antisense strategies aim to prevent the synthesis of a given protein, either by inducing the degradation of its messenger RNA by an RNAse (catabolism) or by physically blocking this single-stranded RNA by forming a partially double-stranded, non-functional molecule.

In the first case, an oligodeoxyribonucleotide binds to the messenger RNA, forming a heteroduplex recognized by RNAse H, which then cleaves the mRNA strand at specific points.

In the second case, there are four possible combinations.

About the target area:

– oligonucleotides targeting the region of the start codon that disrupts the binding of the translation machinery;

– oligonucleotides targeting splicing regions (intron/exon).

About oligonucleotide structures:

– PNA (Peptide Nucleic Acid) oligonucleotides, in which the phosphodiester backbone is replaced by a pseudo-peptide chain binding purine and pyrimidine bases;

– morpholino oligonucleotides, which are more resistant to nucleases.

For these antisense strategies, oligonucleotides are synthetic. Antisense oligonucleotide inhibition of mRNAs requires these oligonucleotides to be chemically modified to improve their binding affinity, biostability and overall pharmacokinetic properties. These molecules can thus neutralize a given messenger RNA under the best possible conditions.

This strategy was initially used to combat AIDS and certain cancers. There is now a great deal of research being carried out in this field to combat parasitosis, inflammatory diseases and cardiovascular disorders.

Fomivirsen is an antisense antiviral drug used in the treatment of cytomegalovirus retinitis. It was the first drug in this class to be approved in the United States, in 1998.

Mipomersen was approved for the US market in 2013. It is an antisense agent used to treat familial hypercholesterolemia, which can lead to myocardial infarction before the age of 30. The drug is not approved in Europe due to long-term liver and heart toxicity.

This therapeutic was recently approved by the FDA and EMA to treat Amyotrophic Lateral Sclerosis by targeting the Superoxide Dismutase 1 (SOD1), with Tofersen denomination.

It should be noted that antisense strategies are under development for enzymes such as telomerase, lipoxygenase, plasma kallikrein, ribonucleotide reductase and thymidylate synthase, to name but a few.

5.4. References

Broutin, M. and Watier, H. (2016). Les biomédicaments 2e partie : les anticorps thérapeutiques. *APBG Biologie Géologie*, 2.

Carroll, D. (2011). Genome engineering with zinc-finger nucleases. *Genetics*, 188, 773–782.

Chandrasegaran, S. and Carroll, D. (2016). Origins of programmable nucleases for genome engineering. *J. Mol. Biol.*, 428(5B), 963–989.

CNRS (2017). Sagasciences [Online]. Available at: http://sagascience.cnrs.fr/dosnano/decouv/vecto/vecto.htm.

Dion, S., Demattéi, M.V., Renault, S. (2007). Les domaines à doigts de zinc. Vers la modification de la structure et de l’activité des génomes. *Med. Sci. (Paris)*, 23(10), 834–839.

Duléry, V. (2007). Développement de méthodes combinatoires pour la découverte de ligands à base de cyclopeptides. PhD Thesis, Université Joseph Fourier, Grenoble.

Dupret, B. and Angrand, P.O. (2014). L’ingénierie des génomes par les TALEN. *Méd. Sci.*, 30, 186–193.

Flinn, A.M. and Gennery, A.R. (2018). Adenosine deaminase deficiency: A review. *Orphanet J. Rare Disease*, 13(1), 65.

Hélène, C. and Saison-Behmoaras, E. (1994). La stratégie antisens : nouvelles approches thérapeutiques. *Médecine/Sciences*, 10, 253–273.

Inserm (2018). Thérapie génique. Une recherche de longue haleine qui porte ses fruits [Online]. Available at: https://www.inserm.fr/information-en-sante/dossiers-information/therapie-genique.

Ishino, Y., Shinagawa, H., Makino, K., Amemura, M. and Nakata, A. (1987) Nucleotide sequence of the iap gene, responsible for alkaline phosphatase isozyme conversion in Escherichia coli, and identification of the gene product. *Journal of Bacteriology*, 169, 5429–5433.

Jinek, M., Chylinski, K., Fonfara, I., Hauer, M., Doudna, J.A., Charpentier, E. (2012). A programmable dual-RNA-guided DNA endonuclease in adaptive bacterial immunity. *Science*, 337(6096), 816–821.

Lino, C.A., Harper, J.C., Carney, J.P., Timlin, J.A. (2018). Delivering CRISPR: A review of the challenges and approaches. *Drug Deliv.*, 25(1), 1234–1257.

Mojica F.J.M., Ferrer C., Juez G., Rodríguez-Valera F. (1995). Long stretches of short tandem repeats are present in the largest replicons of the *Archaea Haloferax mediterranei and Haloferax volcanii* and could be involved in replicon partitioning. *Mol. Microbiol.*, 17, 85–93.

Zhou, Z., Austin, G.L., Shaffer, R., Amstrong, D.D., Gentry, M.S. (2019). Antibody-mediated enzyme therapeutics and applications in glycogen storage diseases. *Trends Mol. Med.*, 25(12), 1094–1109.

Conclusion

Enzymes work in minute doses, remain intact at the end of the reaction and do not alter chemical equilibria. They are the perfect catalysts of the living world. In the eukaryotic cell, where each organelle has specific roles and functions, enzymes are found in the compartment where they are needed, and represent the essential links in catalyzing biological reactions. Enzymes embody the dynamics of life.

When their function is altered in a tissue, an organ or even an organism, pathologies can occur. When pathogenic microorganisms enter the body, their enzymes can be targeted to destroy them or regulate their population.

This book introduces the concepts of enzymatic kinetics, which are essential for the qualitative and quantitative characterization of effectors.

New avenues of research for molecules acting differently on known targets have been detailed, in view of the current context of research concerning inhibitors, activators and other molecules approved by the EMA (European Medicine Agency) and the FDA (Food and Drug Administration).

Since Spallanzani's experiments in the 18th century, considerable progress has been made. Medicinal enzymology, however, has focused on a preferred mode of action on therapeutic targets: competitive inhibition, for pragmatic reasons of simplicity and the need to obtain rapid results.

Over the last 30 years or so, research fields have expanded to diversify modes of inhibition and activation, or to find new molecules with new mechanisms of action. The reasons for this evolution have been outlined. The advent of the molecular modeling era, cutting-edge techniques in molecular biology and organic chemistry, as well as the mastery of the interactome are major factors in the success of this new enzymology research.

In 2020, an unprecedented health crisis caused by the SARS-CoV-2 virus shook up our societies, the pharmaceutical world and the world of research. The established facts underline our populations' lack of scientific knowledge of pathophysiology in general and means of preventing infectious diseases in particular. It would be interesting to have a short general health training course in all branches of education. The basics of health should be part of general culture. Well-informed people are the driving force behind the health successes of our societies. Bridges between disciplines should be developed, so as to ensure the widest possible knowledge base. The 21st century, unlike its predecessors, is no longer one of discovery, but of innovation. The spirit of research and contradiction should find its place again, while respecting the individual. Enzymology, like all the other disciplines of biology and biochemistry, must find its rightful place.

With regard to enzymology, it should be noted that the 2018 Nobel Prize in Chemistry was awarded to work carried out by three researchers directly or indirectly involved with enzymes. George Smith, from the University of Missouri (USA) and Gregory Winter, from the MRC Laboratory (Molecular Biology Laboratory, UK), were rewarded for their research. They have developed the phage display technique, which uses bacteria infected with bacterial viruses (phages) to identify new molecular interactions (antibody/antigen, ligand/receptor and, for our purposes, effector/enzyme). The technique is based on the specific production of a protein of interest by the phage, which may or may not bind to a partner that is immobilized on a microplate well. The phage display technique has enabled the development of new drugs, such as the antibody Humira (Adalimumab), the best-selling drug in 2016. It is used to treat rheumatoid arthritis and other autoimmune diseases.

The other part of the work recognized by this award concerns Frances Arnold, from the California Institute of Technology (USA). She was the first to obtain enzymes that could direct their evolution. This approach has made it possible to generate new catalysts, including enzymes, for the synthesis of new drugs or biofuels, for example. These new enzymes can carry out chemical reactions and make bonds that were previously unknown in the world of living organisms.

The year 2020 was marked by the award of the Nobel Prize in Chemistry to Jennifer Doudna and Emmanuelle Charpentier for their work on CRISPR-Cas9, a system with an enzyme at its heart. The shared history of these two biochemist-geneticists in this field began some 15 years ago, when Emmanuelle Charpentier's work on the immune system of bacteria, and their ability to defend themselves against viruses by conserving viral genetic material, led to the discovery of the role of the Cas9 enzyme. Their results were published in 2012[1]. Since this publication, protocols based on the use of the CRISPR-Cas9 system have revolutionized research, particularly in the field of new drug discovery in genetics (oncology, virology, etc.) as indicated previously in Chapter 5.

The importance of enzyme biocatalysts from the living world has paved the way for the development of asymmetric organocatalysis, which led to the 2021 Chemistry Nobel Prize being awarded to Benjamin List (Max-Planck Institute, Müllheim, Germany) and David MacMillan (Princeton University, USA). This method uses small chiral molecules, whose active part is the active site of an enzyme. It enables a stereoselective catalytic reaction with predominant formation of the expected chiral molecule.

With the discovery of new relationships between enzyme dysfunction and the corresponding disease, there is no doubt that the introduction of new drugs will enrich therapeutic means for the greater good of mankind. These are exciting times for medicinal enzymology, with new areas of research to be developed.

Additional references by the authors

These references by the authors are from the collected volume *Enzymologie : de la structure à la fonction*. They are only available in French.

Dumond, J. and Kirkiacharian, S. (2022a). Enzymologie moléculaire : la catalyse enzymatique. *Techniques de l'ingénieur*.

Dumond, J. and Kirkiacharian, S. (2022b). Cinétique enzymatique : dynamique de transformation. *Techniques de l'ingénieur*.

Dumond, J. and Kirkiacharian, S. (2022c). Enzymologie expérimentale : de la production aux essais. *Techniques de l'ingénieur*.

Dumond, J. and Kirkiacharian, S. (2022d). Enzymologie fonctionnelle : classification des enzymes. Réplication et expression des génomes. *Techniques de l'ingénieur*.

Dumond, J. and Kirkiacharian, S. (2023a). Enzymologie fonctionnelle : enzymes virales et bactériennes. *Techniques de l'ingénieur*.

1 Jinek, M., Chylinski, K., Fonfara, I., Hauer, M., Doudna, J.A., Charpentier, E.A. (2012). Programmable dual-RNA-guided DNA endonuclease in adaptive bacterial immunity. *Science*, 17 August, 337(6096), 816–821.

Dumond, J. and Kirkiacharian, S. (2023b). Enzymologie fonctionnelle : enzymes humaines. *Techniques de l'ingénieur*.

Kirkiacharian, S. and Dumond, J. (2022). Enzymologie structurale. *Techniques de l'ingénieur*.

Kirkiacharian, S. and Dumond, J. (2023). Enzymologie fonctionnelle : pathologies humaines impliquant des enzymes et notion d'enzymatome. *Techniques de l'ingénieur*.

Index

B, C

D, E

F, G

H, I

K, L

M, N

O, P

Q, R

S, T

U, V

W, X, Z

Other titles from

iSTE

in

Chemistry

2020

GODDARD Jean-Philippe, MALACRIA Max, OLLIVIER Cyril
Biphasic Chemistry and the Solvent Case
(Eco-compatibility of Organic Synthesis Set – Volume 3)

2019

GODDARD Jean-Philippe, MALACRIA Max, OLLIVIER Cyril
Multi-component Reactions in Molecular Diversity
(Eco-compatibility of Organic Synthesis Set – Volume 1)
Activation Methods: Sonochemistry and High Pressure
(Eco-compatibility of Organic Synthesis Set – Volume 2)

PIVA Olivier
Retrosynthetic Analysis and Synthesis of Natural Products 1: Synthetic Methods and Applications

2017

VALLS Robert
Inorganic Chemistry: From Periodic Classification to Crystals

Printed and bound by CPI Group (UK) Ltd, Croydon, CR0 4YY
16/09/2026
14963082-0002